D0876347

Introduction to the
Theory of Matroids

Modern Analytic *and* Computational Methods *in* Science *and* Mathematics

A GROUP OF MONOGRAPHS
AND ADVANCED TEXTBOOKS

Richard Bellman, EDITOR
University of Southern California

Published

1. R. E. Bellman, R. E. Kalaba, and Marcia C. Prestrud, Invariant Imbedding and Radiative Transfer in Slabs of Finite Thickness, 1963

2. R. E. Bellman, Harriet H. Kagiwada, R. E. Kalaba, and Marcia C. Prestrud, Invariant Imbedding and Time-Dependent Transport Processes, 1964

3. R. E. Bellman and R. E. Kalaba, Quasilinearization and Nonlinear Boundary-Value Problems, 1965

4. R. E. Bellman, R. E. Kalaba, and Jo Ann Lockett, Numerical Inversion of the Laplace Transform: Applications to Biology, Economics, Engineering, and Physics, 1966

5. S. G. Mikhlin and K. L. Smolitskiy, Approximate Methods for Solution of Differential and Integral Equations, 1967

6. R. N. Adams and E. D. Denman, Wave Propagation and Turbulent Media, 1966

7. R. L. Stratonovich, Conditional Markov Processes and Their Application to the Theory of Optimal Control, 1968

8. A. G. Ivakhnenko and V. G. Lapa, Cybernetics and Forecasting Techniques, 1967

9. G. A. Chebotarev, Analytical and Numerical Methods of Celestial Mechanics, 1967

10. S. F. Feshchenko, N. I. Shkil', and L. D. Nikolenko, Asymptopic Methods in the Theory of Linear Differential Equations, 1967

11. A. G. Butkovskiy, Distributed Control Systems, 1969

12. R. E. Larson, State Increment Dynamic Programming, 1968

13. J. Kowalik and M. R. Osborne, Methods for Unconstrained Optimization Problems, 1968

14. S. J. Yakowitz, Mathematics of Adaptive Control Processes, 1969

15. S. K. Srinivasan, Stochastic Theory and Cascade Processes, 1969

16. D. U. von Rosenberg, Methods for the Numerical Solution of Partial Differential Equations, 1969

17. R. B. Banerji, Theory of Problem Solving: An Approach to Artificial Intelligence, 1969

18. R. Lattès and J.-L. Lions, The Method of Quasi-Reversibility: Applications to Partial Differential Equations. Translated from the French edition and edited by Richard Bellman, 1969

19. D. G. B. Edelen, Nonlocal Variations and Local Invariance of Fields, 1969

20. J. R. Radbill and G. A. McCue, Quasilinearization and Nonlinear Problems in Fluid and Orbital Mechanics, 1970

21. W. Squire, Integration for Engineers and Scientists, 1970

23. T. Hacker, Flight Stability and Control, 1970

24. D. H. Jacobson and D. Q. Mayne, Differential Dynamic Programming, 1970

25. H. Mine and S. Osaki, Markovian Decision Processes, 1970

26. W. Sierpiński, 250 Problems in Elementary Number Theory, 1970

27. E. D. Denman, Coupled Modes in Plasmas, Elastic Media, and Parametric Amplifiers, 1970

28. F. A. Northover, Applied Diffraction Theory, 1971

29. G. A. Phillipson, Identification of Distributed Systems, 1971

30. D. H. Moore, Heaviside Operational Calculus: An Elementary Foundation, 1971

32. V. F. Demyanov and A. M. Rubinov, Approximate Methods in Optimization Problems, 1970

34. C. J. Mode, Multitype Branching Processes: Theory and Applications, 1971

37. W. T. Tutte, Introduction to the Theory of Matroids, 1971

In Preparation

22. T. Parthasarathy and T. E. S. Raghavan, Some Topics in Two-Person Games

31. S. M. Roberts and J. S. Shipman, Two-Point Boundary Value Problems: Shooting Methods

33. S. K. Srinivasan and R. Vasudevan, Introduction to Random Differential Equations and Their Applications

35. R. Tomović and M. Vukobratović, General Sensitivity Theory

36. J. G. Krzyż, Problems in Complex Variable Theory

38. B. W. Rust and W. R. Burrus, Mathematical Programming and the Numerical Solution of Linear Equations

39. N. Buras, Scientific Allocation of Water Resources: Water Resources Development and Utilization—A Rational Approach

Introduction
to the
Theory
of
Matroids

W. T. TUTTE
Faculty of Mathematics
University of Waterloo
Waterloo, Ontario, Canada

CONSULTANT
The Rand Corporation

American Elsevier
Publishing Company, Inc.
NEW YORK · 1971

AMERICAN ELSEVIER PUBLISHING COMPANY, INC.
52 Vanderbilt Avenue, New York, N.Y. 10017

ELSEVIER PUBLISHING COMPANY, LTD.
Barking, Essex, England

ELSEVIER PUBLISHING COMPANY
335 Jan Van Galenstraat, P.O. Box 211
Amsterdam, The Netherlands

International Standard Book Number 0-444-00096-8

Library of Congress Card Number 77-135060

AMS 1970 Subject Classification 05B35

Manufactured in the United States of America

Contents

Foreword

As part of its Air Force Project Rand research program, The Rand Corporation engages in supporting studies in mathematics. The contents of this Report describe the fundamental concepts and methods of matroid theory, as presented in a series of ten lectures by the author at Rand in the summer of 1965.

Matroid theory began with Hassler Whitney's basic paper "On the Abstract Properties of Linear Dependence," *American Journal of Mathematics* (1935), and has since been developed most intensively by the present author, W. T. Tutte, who has obtained deep results in the theory. As the name suggests, a matroid is something like a matrix. The concept in fact generalizes that of "matrix"; in particular, a matroid may be regarded as a generalization of a graph or network. (It is this latter point of view which prevails throughout this Report.) But matroid theory is not just a case of abstraction for abstraction's sake. The subject is already rich in concrete combinatorial applications to circuit theory, network-flow theory, linear and integer programming, and the theory of (0, 1)-matrices, for example, and promises to become more so.

D. R. Fulkerson
Santa Monica, California
Summer, 1970

Preface

This book presents the basic concepts and methods of matroid theory as they appear to the author.

Chapter 1 begins by defining a matroid axiomatically. It then introduces the matroids associated with the structures of graphs and chain-groups.

In Chapter 2 we discuss the subgraphs and contractions of a graph and exhibit corresponding simplifications of chain-groups and matroids. We also study the rank of a matroid.

In Chapter 3 we study a property of matroids called "connection" and show that it corresponds to the property of nonseparability for graphs.

Chapter 4 treats the detailed structure of a matroid, that is, it studies the relation between a given circuit and the rest of the matroid.

In the fifth chapter we consider the regular matroids and their associated chain-groups. The regular matroids mark an interesting half-way stage between the matroids corresponding to graphs on the one hand, and the binary matroids, corresponding to chain-groups over GF(2), on the other.

Chapter 6 is supplementary. It is meant to give very short descriptions of some parts of matroid theory that are not dealt with in the other five chapters. In particular it is concerned with the "Homotopy Theorem" and the characterization of regular and graphic matroids. The author has been informed that his treatment of these matters in his papers on matroids is exceptionally obscure. He hopes that a perusal of Chapter 6 may make it easier to read the detailed proofs.

CHAPTER 1

The Circuits of a Matroid

1.1. INTRODUCTION

A matroid is a structure defined on a finite set E. Several equivalent axiomatric definitions of a matroid are given in the pioneering paper of Hassler Whitney [4]. We use the definition in terms of "circuits."

Consider a class Q of non-null subsets of E. We say that the members of Q are the *circuits* of a *matroid* M on E if the following two axioms hold.

Axiom I.   *No member of Q is a proper subset of another.*

Axiom II.   *Let a and b be distinct members of E. Let X and Y be members of Q such that $a \in X \cap Y$ and $b \in X - Y$. Then there exists $Z \in Q$ such that $b \in Z \subseteq (X \cup Y) - \{a\}$.*

If M is a matroid on E it is convenient to refer to the members of E as the *cells* of M.

Our first two theorems are as follows.

1.11.   Let L be a class of non-null subsets of E. Suppose L satisfies Axiom II. Then if $a \in X \in L$ there is a minimal member Y of L such that $a \in Y \subseteq X$.

By a "minimal" member of L we mean a member that does not contain another.

Proof.   There exists $Y \in L$ such that $a \in Y \subseteq X$. (For example, we may have $Y = X$.) Choose such a Y so that $|Y|$, the number of elements of Y, is as small as possible.

Suppose Y is not minimal in L. Then there exists $Z \in L$ such that $Z \subseteq Y - \{a\}$, by the definition of Y. Choose $b \in Z$. Since L satisfies

1

Axiom II there exists $Z' \in L$ such that $a \in Z' \subseteq (Y \cup Z) - \{b\}$. But this implies $a \in Z' \subset Y$, contrary to the choice of Y.

We conclude that Y is minimal in L. The theorem follows.

1.12. Let L be a class of non-null subsets of E. Suppose L satisfies Axiom II. Then the minimal members of L are the circuits of a matroid M on E.

Proof. Let Q be the class of minimal members of L. Evidently Q satisfies Axiom *I*.

Now let a and b be elements of E, and X and Y elements of Q, such that $a \in X \cap Y$ and $b \in X - Y$. Since L satisfies Axiom II there is a member W of L such that $b \in W \subseteq (X \cup Y) - \{a\}$. By 1.11 there exists $Z \in Q$ such that $b \in Z \subseteq W$. We then have $b \in Z \subseteq (X \cup Y) - \{a\}$. Thus Q satisfies Axiom II.

1.2. CHAIN-GROUPS

Let R be a commutative ring with a unit element and no divisors of zero. Actually in our applications R will be either the ring of integers or the field of residues mod 2.

Let E be a finite set. A *chain* on E over R is a mapping f of E into R. If $a \in E$ we refer to the element $f(a)$ of R as the *coefficient* of a in f. We write $\|f\|$ for the class of all members of E having nonzero coefficients in f. It is the "domain" or "support" of f. The chain on E in which every coefficient is zero is called the zero chain on E and is denoted in formulas by the symbol 0.

Let f and g be chains on E over R. We define their *sum* $f + g$ as the chain on E over R satisfying

$$(f + g)(x) = f(x) + g(x)$$

for each $x \in E$. Again, let f be a chain on E over R and let λ be an element of R. We define the *product* λf as the chain on E over R satisfying

$$(\lambda f)(x) = \lambda f(x)$$

for each $x \in E$.

A *chain-group on E over R* is a class N of chains on E over R that is closed under the operations of addition and multiplication by an element of R.

Let N be a chain-group on E over R. A chain $f \in N$ is called an *elementary chain* of N if it is nonzero and if there is no nonzero chain g of N such that $\|g\|$ is a proper subset of $\|f\|$.

1.21. Let N be a chain-group on E over R. Let f be an elementary chain of N, and let g be a chain of N such that $\|g\| \subseteq \|f\|$. Then either $g = 0$ or there are nonzero elements λ and μ of R such that $\lambda f = \mu g$.

Proof. If $g \neq 0$, choose $a \in \|g\|$. Write $\lambda = g(a)$ and $\mu = f(a)$. Then $\lambda f - \mu g$ is a chain of N and its support is a proper subset of $\|f\|$. Hence $\lambda f - \mu g = 0$, since f is elementary.

1.22. Let N be a chain-group on E over R. Let Q be the class of supports of elementary chains of N. Then Q is the class of circuits of a matroid $M(N)$ on E.

Proof. Let L be the class of supports of nonzero chains of N.

Let X and Y be members of L, and a and b elements of E such that $a \in X \cap Y$ and $b \in X - Y$. There are chains f and g of N such that $X = \|f\|$ and $Y = \|g\|$. Write $h = g(a)f - f(a)g$. Then h is a nonzero chain of N, since $b \in \|h\|$. Hence $\|h\|$ is a member of L satisfying

$$b \in \|h\| \subseteq (X \cup Y) - \{a\}$$

We deduce that L satisfies Axiom II. But Q is the class of minimal members of L. Hence Q is the class of circuits of a matroid $M(N)$, by 1.12.

We refer to $M(N)$ as the *matroid of the chain-group* N.

We define a *binary* chain-group as a chain-group over the field of residues mod 2. A *binary matroid* is the matroid of a binary chain-group.

An *integral* chain-group is a chain-group over the ring of integers.

Let N be an integral chain-group. We define a *primitive* chain of N as an elementary chain of N whose coefficients are restricted to the values 1, −1, and 0. A *regular* chain-group is an integral chain-group in which every elementary chain is an integral multiple of a primitive chain. A *regular* matroid is the matroid of a regular chain-group.

Let f and g be chains of an integral chain-group N. We say f *conforms* to g if $f(a)g(a) > 0$ whenever $f(a) \neq 0$.

1.23. Let N be a regular chain-group on E, and let f be a nonzero chain of N. Then there exists a primitive chain of N conforming to f.

Proof. If possible choose f so that the theorem fails and $\|f\|$ has the least number of elements consistent with this condition.

Since N is regular it has a primitive chain g such that $\|g\| \subseteq \|f\|$. Choose $a \in \|g\|$ so that $f(a)$ has the least possible absolute value. Since $-g$ is primitive we may suppose $g(a)$ and $f(a)$ to have the same sign. Write

$$h = f - f(a)g(a)g.$$

If $h = 0$, then f is a positive multiple of g, and so g conforms to f. If $h \neq 0$ it is clear that h conforms to f. But then $\|h\|$ has fewer elements than $\|f\|$ and so there is a primitive chain k of N conforming to h. But then k conforms to f. In each case the definition of f is contradicted. The theorem follows.

1.24. Let N be a regular chain-group on E, and let f be a nonzero chain on N. Then f can be represented as a sum of primitive chains of N, each conforming to f.

Proof. For each $k \in N$, let $Z(k)$ denote the sum of the absolute values of the coefficients in k.

If possible, choose f so that the theorem fails and $Z(f)$ has the least value consistent with this. By 1.23 there is a primitive chain g of N conforming to f.

Write $h = f - g$. If $h = 0$, then $f = g$. If $h \neq 0$, then it is clear that $Z(h) < Z(f)$ and that h conforms to f. Then h is a sum of primitive chains of N each conforming to h and therefore to f. In each case we find that f satisfies the theorem, contrary to its definition. We conclude that the theorem is true.

Let f and g be integral chains on E, and let q be an integer not less than 2. We call g a *q-representative* of f if the following two conditions are satisfied for each $a \in E$:

$$(i) \ g(a) \equiv f(a) \qquad \mod q \ ,$$

$$(ii) \ |g(a)| < q \ .$$

1.25. Let f be a chain of a regular chain-group N on E. Then for each integer $q \geq 2$, some chain of N is a q-representative of f.

Proof. We note that f is a chain g satisfying condition (i). For each chain g of N satisfying this condition we write $Y(g)$ for the number of elements a of E satisfying $|g(a)| \geq q$. Choose such a g so that $Y(g)$ has the least possible value.

If $Y(g) > 0$, choose $b \in E$ so that $|g(b)| \geq q$. By 1.24 there is a primitive chain h of N that conforms to g and satisfies $h(b) = \pm 1$. Write

$$g_1 = g - qh \ .$$

Clearly g_1 satisfies (i). Moreover we have

$$|g_1(b)| < |g(b)| \ ,$$

$$(|g(a)| < q) \rightarrow (|g_1(a)| < q) \ .$$

Hence $Y(g_1) \leq Y(g)$ with equality only if $|g_1(b)| \geq q$.

If $|g_1(b)| \geq q$ we repeat the process with g_1 replacing g and with the same choice of b. Continuing in this way we eventually obtain a chain g' of N satisfying (i) and such that $Y(g') < Y(g)$. But this, contradicts the definition of g. We conclude that $Y(g) = 0$, that is, g is a q-representative of f.

We can now establish the following theorem about matroids.

1.26. Every regular matroid is binary.

Proof. Let M be a regular matroid. Then $M = M(N)$ for some regular chain-group N. For each $f \in N$ let f' denote the chain over $GF(2)$

derived from it by replacing each coefficient by its residue mod 2. Then the chains f' constitute a binary chain-group N'.

Let X be a circuit of M. There is a primitive chain g of N such that $X = \|g\|$. By the definition of a primitive chain we have $X = \|g'\|$. Hence there is a circuit Y of $M(N')$ such that $Y \subseteq X$.

Conversely, let Y be a circuit of $M(N')$. There is a chain f of N such that $Y = \|f'\|$. But by 1.25 there is a chain g of N that is a 2-representative of f, and we have $\|g\| = \|f'\| = Y$. Hence there is a circuit X of $M = M(N)$ such that $X \subseteq Y$.

Since Axiom I holds for both M and $M(N')$, it follows from the preceding results that these matroids have the same circuits. Hence $M = M(N')$, and the theorem is proved.

1.3. GRAPHS

Let G be a graph, $E(G)$ its set of edges, and $V(G)$ its set of vertices. The two ends of a given edge may be distinct or coincident. In the first case the edge is a *link,* in the second a *loop.*

We *orient* G by distinguishing one end of each edge as positive and the other as negative. In the case of a loop the positive and negative ends coincide. For each $A \in E(G)$ and each $x \in V(G)$ we define an integer $\eta(A, x)$ as follows. If A and x are not incident, or if A is a loop, then $\eta(A, x) = 0$. Otherwise $\eta(A, x) = 1$ or -1 according as x is the positive or the negative end of A.

Chains on $V(G)$ or $E(G)$ over the ring R are called 0-*chains* and 1-*chains,* respectively, of G over R.

With each 1-chain f of G over R there is associated a 0-chain ∂f of G over R defined as follows:

$$(\partial f)(x) = \sum_{A \in E(G)} \eta(A, x) f(A)$$

for each $x \in V(G)$. We call ∂f the *boundary* of f. We call f a *cycle* of G over R if $\partial f = 0$.

The zero chain on $E(G)$ over R is clearly a cycle of G. Moreover it follows from the above definition that boundaries obey the laws

$$\partial(f + g) = \partial f + \partial g,$$

$$\partial(\lambda f) = \lambda \partial f \quad (\lambda \in R).$$

Hence the cycles of G over R constitute a chain-group $\Gamma_R(G)$ on E over R. We call this the *cycle-group* of G over R. A *polygon* of G is a subgraph defined by k edges $A_1, A_2, \ldots, A_k$ and the same number of vertices $a_1, a_2, \ldots, a_k = a_0$, subject to the conditions that $k \geq 1$ and that the two ends of A_i are a_{i-1} and $a_i (1 \leq i \leq k)$. A polygon with exactly k edges is often called a *k-gon*. Thus a 1-gon consists solely of a loop and its incident vertex, and a k-gon with $k > 1$ has no loop.

Given a polygon K of G we define a 1-cycle f_K over R as follows. If $A \in E(G) - E(K)$, then $f_K(A) = 0$. Otherwise we use the preceding notation and write $f_K(A_i) = \eta(A_i, a_i) = 1, -1$, or 0 according as a_i is the positive end of A_i, the negative end of A_i, or both. Thus f_K depends on one specified enumeration of the edges and vertices of K. We refer to it as "the cycle associated with K."

If S is a non-null proper subset of $E(K)$, it follows from the definition of a polygon that some vertex v of G is incident with exactly one edge of S. Then v has a nonzero coefficient in the boundary of any 1-chain of G over R whose support is S. We deduce that f_K is an elementary chain of $\Gamma_R(G)$. Moreover, if R is the ring I of integers, then f_K is a primitive chain of $\Gamma_I(G)$.

Now let f be any elementary chain of $\Gamma_R(G)$. Let H be the subgraph of G consisting of the edges in $\|f\|$ and their incident vertices. By the definition of a cycle, each vertex of H is incident with at least two edges in H. Hence, by an easy exercise in graph theory, there is a polygon K that is a subgraph of H. Since f is elementary we have $\|f_K\| = \|f\|$. Choose $a \in f$ and write $\lambda = f(a)f_K(a)$. Then $f = \lambda f_K$, since $\|f - \lambda f_K\|$ is a proper subset of $\|f\|$.

From the foregoing observations we deduce the following two theorems.

1.31. Let S be a subset of $E(G)$. Then in order that there shall exist an elementary chain g of $\Gamma_R(G)$ such that $S = \|g\|$, it is necessary and sufficient that S shall be the set of edges of a polygon of G.

1.32. $\Gamma_I(G)$ is a regular chain-group.

To each 0-chain f of G over R there corresponds a 1-chain δf of G over R called the *coboundary* of f. It is defined as follows:

$$(\delta f)(A) = \sum_{x \in V(G)} \eta(A, x) f(x)$$

for each $A \in E(G)$. Thus if p is the positive and q the negative end of A we have

$$(\delta f)(A) = f(p) - f(q).$$

We note that $\delta 0 = 0$. Moreover, coboundaries satisfy the laws

$$\delta(f + g) = \delta f + \delta g,$$

$$\delta(\lambda f) = \lambda \delta(f).$$

Hence the coboundaries of the 0-chains of G over R are the elements of a chain-group $\Delta_R(G)$ on $E(G)$ over R. We call $\Delta_R(G)$ the *coboundary group* of the oriented graph G, over R.

Let S be a non-null set of edges of G, and let H be the graph obtained from G by deleting the edges of S but retaining all the vertices. It may happen that two components H_1 and H_2 of H are such that each edge of S has one end (in G) in $V(H_1)$ and the other in $V(H_2)$. If so, we say that S is a *bond* of G, and that H_1 and H_2 are the two *end-graphs* of S in G.

Given a bond S of G with end-graphs H_1 and H_2, we can define a 0-chain k_S of G over R as follows: $k_S(x) = 1$ if $x \in V(H_1)$, and $k_S(x) = 0$ otherwise. We now write $f_S = \delta k_S$, and refer to f_S as "the coboundary associated with S." Clearly $\|f_S\| = S$. Moreover the coefficients in f_S are restricted to the values 0, 1, and -1. The chain f_S is dependent on the order in which the two end-graphs are taken; interchanging H_1 and H_2 replaces f_S by $-f_S$.

Suppose there is a nonzero coboundary $g = \delta h$ of G over R such that $\|g\|$ is a proper subset of $\|f_S\|$. Choose $A \in \|f_S\| - \|g\|$. Since the edges of H_1 and H_2 have zero coefficients in g, it follows that all the vertices of H_1 have the same coefficient t_1 in h, and all the vertices of h_2 have the same coefficient t_2 in h. Moreover $t_1 = t_2$, since $g(A) = 0$. But then $g(B) = 0$ for each $B \in S$, contrary to the definition of g. We deduce that f_S is an elementary chain of $\Delta_R(G)$. If $R = I$, it is a primitive chain by the restriction on the values of its coefficients.

Let $g = \delta h$ be any elementary chain of $\Delta_R(G)$. Then there is an edge A of G whose positive and negative ends have different coefficients λ and μ in h. Let h_1 be the 0-chain of G such that $h_1(x) = 1$ if $h(x) = \lambda$, and $h_1(x) = 0$ otherwise. Write $g_1 = \delta h_1$. Then $A \in \|g_1\| \subseteq \|g\|$. Accordingly, $\|g_1\| = \|g\|$, since g is elementary. Write $S = \|g\|$.

Let H be derived from G by deleting the edges of S. Let H_1 and H_2 be the components of H containing the positive and negative end of A, respectively. Then the vertices of H_1 must all have coefficient 1 in h_1, and those of H_2 have coefficient 0 in h_1. Now let h_2 be the 0-chain of G over R in which the vertices of H_1 have coefficient 1 and all other vertices have coefficient 0. Then $A \in \|\delta h_2\| \subseteq \|g_1\| = \|g\| = S$. Hence $\|\delta h_2\| = S$, since g is elementary. Thus each edge of S has an end in $V(H_1)$. Similarly, each such edge has an end in $V(H_2)$. We conclude that S is a bond of G.

From the foregoing results we deduce the following two theorems.

1.33. Let S be a subset of $E(G)$. Then in order that there shall exist an elementary chain g of $\Delta_R(G)$ such that $S = \|g\|$, it is necessary and sufficient that S shall be a bond of G.

1.34. $\Delta_I(G)$ is a regular chain-group.

From 1.31 and 1.33 we deduce that the matroids of $\Gamma_R(G)$ and $\Delta_R(G)$ are independent of the ring R. The circuits of the first matroid are defined by the polygons of G, and those of the second by the bonds of G. We observe further that each matroid is independent of the orientation assigned to G. We denote the matroids of $\Gamma_R(G)$ and $\Delta_R(G)$ by $P(G)$ and $B(G)$, respectively. $P(G)$ is the *polygon-matroid* and $B(G)$ the *bond-matroid* of G.

1.4. ORTHOGONALITY

If f and g are two chains on E over R, we write

$$(f \cdot g) = \sum_{x \in E} f(x)g(x).$$

We say that f and g are *orthogonal* if $(f \cdot g) = 0$.

Now let N be a chain-group on E over R. Define N^* as the class of all chains h on E over R such that h is orthogonal to every chain on N. It is easy to verify that N^* is a chain-group on E over R. We call it the *dual* chain-group of N.

The concept of duality can be generalized from chain-groups to matroids. Here we are concerned only with its application to the groups $\Gamma_R(G)$ and $\Delta_R(G)$ of a graph G, with a specified orientation.

1.41. Let k be a 1-chain of G over R. Then k is a cycle of G if and only if it is orthogonal to every coboundary of G over R.

Proof. Let f be the coboundary of a 0-chain g of G over R. Then

$$(k \cdot f) = \sum_{A \in E(G)} k(A)f(A)$$

$$= \sum_{A \in E(G)} \sum_{x \in V(G)} \eta(A, x)k(A)g(x)$$

$$= \sum_{x \in V(G)} g(x) \cdot (\partial k)(x).$$

If k is a cycle we have $\partial k = 0$, and so $(k \cdot f) = 0$ for each $f \in \Delta_R(G)$. Conversely, if k is orthogonal to every $f \in \Delta_R(G)$, then

$$\sum_{x \in V(G)} g(x) \cdot (\partial k)(x) = 0$$

for arbitrary integers $g(x)$. Hence $\partial k = 0$, and so k is a cycle.

Corollary. $\Gamma_R(G) = [\Delta_R(G)]^*$.

1.42. Let k be a 1-chain of G over R. Then K is a coboundary of G if and only if it is orthogonal to every cycle of G over R.

Proof. If k is a coboundary of G, it is orthogonal to every chain of $\Gamma_R(G)$ by 1.41.

Suppose therefore that k is orthogonal to every member of $\Gamma_R(G)$. Enumerate the components of G as $G_1, G_2, \ldots, G_n$. In each component G_i, distinguish one vertex v_i.

Consider an arbitrary vertex x of G. Let it be in the component G_j. Then we can find a sequence $(a_1, a_2, \ldots, a_s)$, of one or more vertices of G_j, with the following properties:

(i) $a_1 = v_i$ and $a_s = x$.
(ii) If $1 \leq r < s$, there is an edge A_r of G_j with ends a_r and a_{r+1}.

We now define a 1-chain f_x as follows: $f_x(A_r) = \eta(A_r, a_r)$, and $f_x(B) = 0$ if B is not of the form A_r. We assume that for each r, $1 \leq r < s$, just one edge is specified as A_r. We may thus write

$$\partial f_x = v_i - x$$

to indicate that v_i has coefficient 1, x coefficient -1, and every other vertex coefficient 0 in ∂f_x, unless $v_i = x$, in which case $\partial f_x = 0$. We now define a 0-chain h as follows:

$$h(x) = (k \cdot f_x), \qquad x \in V(G).$$

We can show that $h(x)$ is independent of the particular choice of f_x. For if g_x is another 1-chain over R such that $\partial g_x = v_i - x$, then $\partial (f_x - g_x) = 0$. Hence $[k \cdot (f_x - g_x)] = 0$ by the definition of k, and therefore $(k \cdot f_x) = (k \cdot g_x)$.

Let A be an edge of G with ends a and b. We can arrange, interchanging a and b if necessary, that A serves as the edge A_{s-1} in the calculation of f_b, with $b = a_s$. Then

$$
\begin{aligned}
(\delta h)(A) &= \eta(A, a)\,[h(a) - h(b)] \\
&= \eta(A, a)\,\{(k \cdot f_a) - (k \cdot f_b)\} \\
&= \eta(A, a)\,[k \cdot (f_a - f_b)] \\
&= -\eta(A, a) \cdot k(A) \cdot \eta(A, a) \\
&= -k(A).
\end{aligned}
$$

Hence $k = \delta(-h)$, $k \in \Delta_R(G)$. This completes the proof of the theorem.

Corollary. $\Delta_R(G) = [\Gamma_R(G)]^*$.

CHAPTER 2

Minors

2.1. SUBGRAPHS AND CONTRACTIONS

Let G be a graph and let S be any subset of $E(G)$. We define the subgraph $G:S$ of G by the rules $V(G:S) = V(G)$ and $E(G:S) = S$. The subgraph $G \cdot S$, called here the *reduction* of G to S, is defined by the edges of S and their incident vertices. Thus $G \cdot S$ is derived from $G:S$ by deleting all the isolated vertices.

If G is given as oriented we suppose $G:S$ and $G \cdot S$ to be correspondingly oriented. That is, each edge of S has to have the same positive and negative ends in $G:S$ and $G \cdot S$ as in G.

Consider the graph $G:[E(G) - S]$. It is possible to define a graph H whose vertices are the components of $G:[E(G) - S]$ and whose edges are the members of S. If $A \in S$, then the ends of A in H are those components of $G:[E(G) - S]$ that contain the ends of A in G. Thus A may be a link in G but a loop in H. We write $H = G$ ctr S and call H the *contraction* of G to S.

If the identities of the vertices of G are not important we may say, less precisely, that G ctr S is formed from G by contracting each component of $G:[E(G) - S]$ to a single vertex. We may also say that G ctr S is formed from G by a series of "elementary contractions" in each of which one edge of $E(G) - S$ is contracted to a single vertex.

By deleting the isolated vertices of G ctr S we obtain from it the *reduced contraction* $G \times S = (G$ ctr $S) \cdot S$ of G to S.

If G is oriented we suppose G ctr S and $G \times S$ to be correspondingly oriented. Thus if $A \in S$, then its positive and negative ends in either graph G ctr S or $G \times S$ contain as vertices the positive and negative ends, respectively, of A in G.

2.2. MINORS OF CHAIN-GROUPS

In this section we find the operations on $\Gamma_R(G)$ and $\Delta_R(G)$ that correspond to taking a reduction $G \cdot S$ or a reduced contraction $G \times S$ of an oriented graph G.

Let N be any chain-group on a set E over R. If $S \subseteq E$, we define the *restriction* to S of a chain f of N as the chain g on S such that $f(x) = g(x)$ for each $x \in S$.

The restrictions to S of the chains of N constitute a chain-group on S. We call this the *reduction* of N to S and denote it by $N \cdot S$. Another chain-group on S is constituted by the restrictions to S of those chains f of N for which $\|f\| \subseteq S$. We call this the *contraction* of N to S and denote it by $N \times S$.

Consider the oriented graph $G \cdot S$. If $f = \delta g$ in G, it is clear that the restriction of f to S is the coboundary, in $G \cdot S$, of the restriction of g to $V(G \cdot S)$. It follows that

2.211. $\Delta_R(G \cdot S) = [\Delta_R(G)] \cdot S.$

If T is a subset of a set E, and h is a chain on T, then there is a chain k on E, over the same ring, such that h is the restriction of k to T and $k(A) = 0$ for each $A \in E - T$. We call k the *zero extension* of h to E.

Returning to the graph $G \cdot S$, we let f be any chain on S over R, and g its zero extension to $E(G)$. It is evident that ∂g is the zero extension of ∂f to $V(G)$. Thus f is a cycle of $G \cdot S$ if and only if g is a cycle of G. We conclude that

2.212. $\Gamma_R(G \cdot S) = [\Gamma_R(G)] \times S.$

Now consider the graph $G \times S$. Let f be a chain on S over R, and g its zero extension to $E(G)$. Suppose $f = \delta h$ in $G \times S$. We now define a 0-chain h_1 of G as follows. Suppose $x \in V(G)$, and let X be the component of $G : [E(G) - S]$ containing it. If $X \in V(G \times S)$, then $h_1(x) = h(X)$. Otherwise $h_1(x) = 0$. It is clear that $\delta h_1 = g$ in G.

Conversely, suppose g is the coboundary in G of a 0-chain h_1. Then in any component of $G : [E(G) - S]$ all the vertices have equal coefficients in g; and if the component represents an isolated vertex of G ctr S, we may suppose these coefficients to be zero. Hence there is a 0-chain h of $G \times S$ such that h and h_1 are related as in the preceding paragraph. Clearly $\delta h = f$ in $G \times S$. We conclude that

2.213. $\Delta_R(G \times S) = [\Delta_R(G)] \times S.$

In the next argument we let f be a chain on $E(G)$ over R, and g its restriction to S. We enumerate the components of $G : [E(G) - S]$ as $X_1, X_2, \ldots, X_s$. We can write

$$f = f_S + \sum_{i=1}^{s} q_i,$$

where $\|f_S\| \subseteq S$, and $\|q_i\| \subseteq E(X_i)$ for each i.

Suppose $X_j \in V(G \times S)$. Its coefficient in ∂g, for the graph $G \times S$, is given by

$$(\partial g)(X_j) = \sum_{x \in V(X_j)} (\partial f_S)(x),$$

where ∂f_S is taken in G.

Suppose $\partial f = 0$ in G. Then

$$-(\partial g)(X_j) = \sum_{x \in V(X_j)} (\partial q_j)(x)$$

$$= 0,$$

for the sum of coefficients in any boundary is zero, by the definition of a boundary. Hence $\partial g = 0$ in $G \times S$.

Conversely, suppose $\partial g = 0$ in $G \times S$. Then, if h_S is the zero extension of g to $E(G)$,

$$\sum_{x \in V(X_i)} (\partial h_S)(x) = 0$$

for each X_i. However, if y and z are any two vertices of X_i, we can find a 1-chain of X_i over R whose boundary is $y - z$, as in the proof of 1.42 with v_i and x. Hence we deduce the existence of a 1-chain p_i on G such that $\|p_i\| \subseteq E(X_i)$ and

$$(\partial p_i)(x) = - (\partial h_S)(x)$$

for each $x \in V(X_i)$. Writing

$$h = h_S + \sum_{i=1}^{s} p_i ,$$

we find that $\partial h = 0$ and that g is the restriction of h to S.

We deduce that

2.214. $\Gamma_R(G \times S) = [\Gamma_R(G)] \cdot S.$

The preceding results are of fundamental importance for relating the theory of chain-groups to that of graphs. We should also note the following identities in the theory of chain-groups. We suppose that N is a chain-group on E over R and that $T \subseteq S \subseteq E$.

2.221. $(N \times S) \times T = N \times T,$

2.222. $(N \cdot S) \cdot T = N \cdot T,$

2.223. $(N \cdot S) \times T = \{N \times [E - (S - T)]\} \cdot T,$

2.224. $(N \times S) \cdot T = \{N \cdot [E - (S - T)]\} \times T.$

The first two identities are immediate from the definitions. To prove 2.223 we observe that each side of the formula represents the class of restrictions to T of those chains of N that have only zero coefficients in $S - T$. We can obtain 2.224 by writing $E - (S - T)$ for S in 2.223.

We refer to a chain-group $(N \cdot S) \times T$ as a *minor* of N. In particular the minors of N include N itself and all its reductions and contractions ($N = (N \cdot E) \times E$, $N \cdot S = (N \cdot S) \times S$, and $N \times S = (N \cdot E) \times S$).

2.23. Every minor of a minor of N is a minor of N.

Proof. A minor of a minor of N is of the form $\{[(N \cdot S) \times T] \cdot U\} \times V$. Using 2.224 we can rewrite this in the form $\{[(N \cdot S) \cdot T_1] \times U\} \times V$, that is, $(N \cdot T_1) \times V$ by 2.221 and 2.222.

2.24. If $R = I$ and N is regular, then every minor of N is regular.

Proof. Suppose $S \subseteq E$. Let f be an elementary chain of $N \times S$. Its zero extension h to E is then an elementary chain of N. Since N is regular, h is a multiple of a primitive chain k of N. Let g be the restriction of k to S. Then $g \in N \times S$, and f is a multiple of g. It follows that g is primitive in $N \times S$. We deduce that $N \times S$ is regular.

Now let f be an elementary chain of $N \cdot S$. It is the restriction to S of a chain h of N, by the definition of $N \cdot S$. There is a primitive chain k of N such that $\|k\| \cap S$ is not null and $\|k\| \subseteq \|h\|$, by 1.24. Let g be the restriction of k to S. Since $\|g\|$ is non-null and a subset of $\|f\|$ we deduce that g is a primitive chain of $N \cdot S$, and that f is one of its multiples. We deduce that $N \cdot S$ is regular. The theorem follows.

2.3. MINORS OF MATROIDS

We extend the definition of minors to matroids as follows.

Let M be a matroid on a set E, and Q its set of circuits. Let S be a subset of E.

Let L be the class of all non-null intersections of S with members of Q. Since Q satisfies Axiom II, it is clear that L does too. Let L' be the set of all minimal members of L. Then L' is the set of circuits of a matroid on S. We denote this matroid by $M \cdot S$ and call it the *reduction* of M to S.

Let Q_1 be the class of all members of Q that are subsets of S. Clearly Q_1 satisfies Axioms I and II. It is the set of circuits of a matroid on S. We denote this matroid by $M \times S$ and call it the *contraction* of M to S.

2.31. Let N be a chain-group on a set E over R. Let S be a subset of E. Then

$$M(N \cdot S) = M(N) \cdot S,$$

$$M(N \times S) = M(N) \times S.$$

This result follows by a comparison of definitions, it being noted that the elementary chains of $N \times S$ are the restrictions to S of those elementary chains of N whose supports are subsets of S.

Let G be any graph. We assign an arbitrary orientation to G and use 2.211–2.214 and 2.31. We obtain the following results:

2.321. $B(G \cdot S) = B(G) \cdot S,$

2.322. $B(G \times S) = B(G) \times S,$

2.323. $P(G \cdot S) = P(G) \times S,$

2.324. $P(G \times S) = P(G) \cdot S.$

Since polygon-matroids and bond-matroids do not depend on orientation we can consider G to be an unoriented graph in these formulas.

Formulas 2.221–2.224 generalize as follows. Let M be a matroid on a set E and suppose $T \subseteq S \subseteq E$. Then

2.331. $(M \times S) \times T = M \times T,$

2.332. $(M \cdot S) \cdot T = M \cdot T,$

2.333. $(M \cdot S) \times T = \{M \times [E - (S - T)]\} \cdot T,$

2.334. $(M \times S) \cdot T = \{M \cdot [E - (S - T)]\} \times T.$

Formula 2.331 follows at once from the definitions.

To prove 2.332 suppose X is a circuit of $(M \cdot S) \cdot T$. Then X is the intersection with T of a circuit of $M \cdot S$, and therefore of a circuit of M. Hence there is a circuit Y of $M \cdot T$ such that $Y \subseteq X$.

Conversely, suppose Y is a circuit of $M \cdot T$. Then there is a circuit Z of M such that $Z \cap T = Y$. But Z contains a circuit Z_1 of $M \cdot S$ which meets Y, by 1.11. Hence there is a circuit X of $(M \cdot S) \cdot T$ such that $X \subseteq Z_1 \cap T \subseteq Z \cap T = Y$.

Applying Axiom I to $(M \cdot S) \cdot T$ and $M \cdot T$ we deduce that these matroids are identical.

We prove 2.333 in a similar way. Let X be a circuit of $(M \cdot S) \times T$. Then X is a circuit of $M \cdot S$ contained in T. It follows that there is a circuit X_1 of M such that $X_1 \cap T = X$, and $X_1 \cap (S - T)$ is null. But then X_1 is a circuit of $M \times [E - (S - T)]$. Hence there is a circuit Y of $\{M \times [E - (S - T)]\} \cdot T$ such that $Y \subseteq X_1 \cap T = X$.

Conversely, suppose Y is a circuit of $\{M \times [E - (S - T)]\} \cdot T$. Then there is a circuit Y_1 of $M \times [E - (S - T)]$ such that $Y = Y_1 \cap T$. But then Y_1 is a circuit of M such that $Y_1 \cap (S - T)$ is null. Hence there is a circuit X of $M \cdot S$ such that $X \subseteq Y_1 \cap S = Y_1 \cap T = Y$. But then $X \subseteq T$, and so X is a circuit of $(M \cdot S) \times T$.

An application of Axiom I completes the proof.

We obtain 2.334 by writing $E - (S - T)$ for S in 2.333.

We refer to a matroid of the form $(M \cdot S) \times T$ as a *minor* of M.

2.34. Every minor of a minor of a matroid M is a minor of M.

The proof is analogous to that of 2.23.

2.35. Let N be any chain-group. Then the minors of $M(N)$ are the matroids of the minors of N (by 2.31).

2.36. The minors of a binary matroid are binary, and those of a regular matroid are regular (by 2.35 and 2.24).

2.4. RANK

Let M be a matroid on a set E. We describe a subset S of E as a *flat* of M if it is a union of circuits of M. The null subset of E is counted as a null union of circuits, and therefore as a flat. For each subset S of E there is a flat $\langle S \rangle$ of M defined as the union of those circuits of M that are contained in S.

Let S be any flat of M. A *descending sequence* of S is a sequence $(S_1, S_2, \ldots, S_k)$ of flats of M such that (i) $S_1 = S$, (ii) $S_k = \phi$, and (iii) each member of the sequence other than S_1 is a proper subset of its predecessor. The number k of terms in the descending sequence is its *length*. The *rank* $r(S)$ of S is one less than the maximum number of terms for a descending sequence of S. From this definition we have

2.41. Let S be any flat of M. Then $|S| \geq r(S) \geq 0$. The second equality holds only if $S = \phi$.

2.42. Let S be any flat of M. Let a and b be cells of S such that $\langle S - \{a\}\rangle$ and $\langle S - \{b\}\rangle$ are distinct. Then $\langle S - \{a, b\}\rangle$ is a proper subset of both $\langle S - \{a\}\rangle$ and $\langle S - \{b\}\rangle$.

Proof. Since $\langle S - \{a\}\rangle$ and $\langle S - \{b\}\rangle$ are distinct we can adjust the notation so that S contains a circuit X of M such that $a \in X$ and $b \notin X$.

There is another circuit Y of M such that $b \in Y \subseteq S$. If $a \in Y$ there is a circuit Y' of M such that $b \in Y' \subseteq S - \{a\}$ by Axiom II. We deduce that there is a circuit Z of M (Y or Y') such that $Z \subseteq S$, $a \notin Z$, and $b \in Z$.

We have $X \subseteq \langle S - \{b\}\rangle$ and $Z \subseteq \langle S - \{a\}\rangle$. But neither X nor Z can be contained in $\langle S - \{a, b\}\rangle$. The theorem follows.

2.43. Let S be any flat of M and suppose $a \in S$. Then

$$r(\langle S - \{a\}\rangle) = r(S) - 1 .$$

Proof. This is trivially true when $r(S) = 0$. It is also true when $r(S) = 1$, for then ϕ is the only flat that is a proper subset of S.

Suppose the theorem true whenever $r(S)$ is less than some integer $q > 1$, and consider the case $r(S) = q$.

Let T be the second member of a descending sequence of S of maximum length $r(S) + 1$. Evidently $r(T) = r(S) - 1$. Choose $b \in S - T$. Then $T \subseteq \langle S - \{b\}\rangle \subset S$, and so $T = \langle S - \{b\}\rangle$ by the definition of T.

If $\langle S - \{a\}\rangle = T$, the theorem is proved. In the remaining case $\langle S - \{a, b\}\rangle$ is a proper subset of both T and $\langle S - \{a\}\rangle$, by 2.42. But it is clear that $\langle S - \{a, b\}\rangle = \langle T - \{a\}\rangle$. Hence $r(\langle S - \{a, b\}\rangle) = r(S) - 2$, by the inductive hypothesis. Since $\langle S - \{a\}\rangle$ properly contains $\langle S - \{a, b\}\rangle$ and is a proper subset of S, we must have $r(\langle S - \{a\}\rangle) = r(S) - 1$.

2.44. If S and T are flats of M such that $S \subset T$, then $r(S) < r(T)$. Moreover, there is a flat U of M such that $S \subset U \subseteq T$ and $r(U) = r(S) + 1$.

Proof. The theorem is trivially true when $r(T) = 0$. Assume it true whenever $r(T)$ is less than some positive integer q, and consider the case $r(T) = q$.

Choose $a \in T - S$. Write $T_1 = \langle T - \{a\} \rangle$. Then $r(T_1) = r(T) - 1$, by 2.43, and $S \subseteq T_1$.

If $S = T_1$ the theorem holds with $U = T$.

If $S \subset T_1$ it follows by the inductive hypothesis that $r(S) < r(T_1) < r(T)$ and that there exists a flat U of M such that $S \subset U \subseteq T_1 \subset T$ and $r(U) = r(S) + 1$. The theorem is thus true when $r(T) = q$. It follows in general by induction.

2.45. Let S and T be flats of M such that $S \subseteq T$. Then there exists a flat U of M such that $U \subseteq T$, $\langle U \cap S \rangle = \phi$, and $r(U) = r(T) - r(S)$.

Proof. If $r(S) = 0$ the theorem holds with $U = T$. Assume it true whenever $r(S)$ is less than some positive integer q and consider the case $r(S) = q$.

Choose $a \in S$. Then $\langle S - \{a\} \rangle \subseteq \langle T - \{a\} \rangle$. By the inductive hypothesis there is a flat U of M such that $U \subseteq \langle T - \{a\} \rangle$, $\langle U \cap \langle S - \{a\} \rangle \rangle = \phi$, and $r(U) = r(\langle T - \{a\} \rangle) - r(\langle S - \{a\} \rangle)$.

We deduce from 2.43 that $r(U) = r(T) - r(S)$. Moreover, $U \subseteq T$, and if $\langle U \cap S \rangle \neq \phi$, there must be a circuit Z of M such that $a \in Z \subseteq S$, and $Z \subseteq U$. This is impossible since $U \subseteq \langle T - \{a\} \rangle$. Hence the theorem holds when $r(S) = q$, and is true in general by induction.

2.46. Let S and T be flats of M. Then

$$r(S \cup T) + r(\langle S \cap T \rangle) \geq r(S) + r(T) .$$

Proof. This is trivially true when $T \subseteq S$. Assume it true when $|T - S|$ is less than some positive integer q, and consider the case $|T - S| = q$.

Choose $a \in T - S$. By the inductive hypothesis,

$$r(S \cup \langle T - \{a\} \rangle) + r(\langle S \cap \langle T - \{a\} \rangle \rangle) \geq r(S) + r(\langle T - \{a\} \rangle)$$

But $S \cup T$ is a flat of M satisfying $S \cup T \supset S \cup \langle T - \{a\} \rangle$. It is also clear that

$$\langle S \cap T \rangle \supseteq \langle S \cap \langle T - \{a\} \rangle \rangle \, .$$

Hence

$$r(S \cup T) + r(\langle S \cap T \rangle) \geq r(S) + r(\langle T - \{a\} \rangle) + 1 \, ,$$

by 2.44,

$$= r(S) + r(T) \, ,$$

by 2.43.

The theorem follows by induction.

2.5. RANKS OF MINORS

Let M be a matroid on a set E. We define the *rank* $r(M)$ of M as the rank of the flat $\langle E \rangle$ of M.

There is another kind of rank, used by Whitney, which is the greatest number of cells in a subset S of E containing no circuit of M. Let us denote this rank by $\rho(M)$. It is not difficult to show that $\rho(M) = |E| - r(M)$.

If $S \subseteq E$, it is clear from the definitions that $r(M \times S)$ is the rank of the flat $\langle S \rangle$ of M.

2.51. Suppose $a \in S \subseteq E$. **If there is a circuit** Z **of** M **such that** $a \in Z \subseteq S$, **then**

$$r[M \times (S - \{a\})] = r(M \times S) - 1 \, ;$$

otherwise

$$r[M \times (S - \{a\})] = r(M \times S) \, .$$

Proof. In the first case, $a \in \langle S \rangle$. Hence $r[M \times (S - \{a\})] = r(\langle S - \{a\} \rangle) = r(\langle S \rangle) - 1 = r(M \times S) - 1$. In the second case, $\langle S - \{a\} \rangle = \langle S \rangle$.

Ranks of reductions can be related to those of contractions as follows.

2.52. Suppose $S \subseteq E$ and $\overline{S} = E - S$. Then

$$r(M \cdot S) + r(M \times \overline{S}) = r(M).$$

Proof. This result is trivial when $|E| = 0$. Assume it true when $|E|$ is less than some positive integer q, and consider the case $|E| = q$.

Suppose we can choose $a \in S$ so that some circuit X of $M \cdot S$ satisfies $a \in X$. Then a belongs also to a circuit of M. We have

$$r(M \cdot S) = r[(M \cdot S) \times (S - \{a\})] + 1,$$

by 2.51,

$$= r[M \times (E - \{a\})] \cdot (S - \{a\}) + 1,$$

by 2.333,

$$= r[M \times (E - \{a\})] - r(M \times \overline{S}) + 1,$$

by 2.331 and the inductive hypothesis; and

$$= r(M) - r(M \times \overline{S}),$$

by 2.51. The theorem follows by induction, for if $M \cdot S$ has no circuit we have $r(M \cdot S) = 0$ and $\langle E \rangle = \langle \overline{S} \rangle$ in M.

2.6. RANKS OF CHAIN-GROUPS

Let N be a chain-group on a set E over a ring R.

Let $f_1, f_2, \ldots, f_k$ be chains of N. We say they are linearly dependent (over R) if there are elements $\lambda_1, \lambda_2, \ldots, \lambda_k$ of R, not all zero, such that

$$\sum_{i=1}^{1} \lambda_i f_i = 0.$$

2.61. Let S be a flat of $M(N)$. Then the maximum number of linearly independent chains f of N such that $\|f\| \subseteq S$ is $r(S)$.

Proof. If possible choose S so that the theorem fails and $r(S)$ has the least value consistent with this. Then $r(S) \geq 1$, for otherwise S is null and the theorem is trivially true.

Choose $a \in S$. Since S is a flat of $M(N)$ there is a chain f of N such that $a \in \|f\| \subseteq S$. But $r(\langle S - \{a\}\rangle) = r(S) - 1$, by 2.43. Hence, by the choice of S we can find $r(S) - 1$ linearly independent chains of N with supports in $\langle S - \{a\}\rangle$. Adjoining f to these we obtain $r(S)$ linearly independent chains of N with supports in S.

We deduce that there are $k = r(S) + 1$ linearly independent chains $f_1, f_2, \ldots, f_k$ of N with supports in S. At most $k - 2$ of them have supports in $\langle S - \{a\}\rangle$. We can adjust the notation so that $a \in \|f_i\|$ if $i \leq s$, and $a \notin \|f_i\|$ if $i > s$, where s is some integer satisfying $2 \leq s \leq k$. For each $i \leq s$ we write

$$g_i = f_1(a) f_i - f_i(a) f_1.$$

By the choice of S, and since $\|g_i\| \subseteq \langle S - \{a\}\rangle$, we can find elements $r_2, r_3, \ldots, r_k$ of R, not all zero, such that

$$\sum_{i=2}^{s} r_i g_i + \sum_{i=s+1}^{k} r_i f_i = 0.$$

Writing the left-hand side of this equation in terms of the k chains f_i we find that these chains are not linearly independent. This contradiction establishes the theorem.

2.7. THE RANKS OF $P(G)$ AND $B(G)$

Let G be any graph. Using 2.212 and 2.43 we find that $r[P(G)]$ is the least number of edges whose deletion from G destroys every polygon. This is $p_1(G)$, the *Betti number of dimension* 1 of G.

The Betti number $p_0(G)$ of dimension 0 of G is defined as the number of components of G. It is shown in graph theory that if $\alpha_0(G)$ and $\alpha_1(G)$ denote the numbers of vertices and edges of G, respectively, then

$$\alpha_1(G) - \alpha_0(G) = p_1(G) - p_0(G).$$

We thus have

2.71. $r[P(G)] = p_1(G) = \alpha_1(G) - \alpha_0(G) + p_0(G).$

Similarly, $r[B(G)]$ is the least number of edges whose contraction, each to a single vertex, destroys every bond. We evaluate it as follows

2.72. $r[B(G)] = \alpha_0(G) - p_0(G).$

Proof. When we contract a single edge we convert each component of the original graph into a component of the contracted one, that is, $p_0(G)$ is invariant under the operation. But $\alpha_0(G)$ is reduced by exactly 1. To destroy every bond we must destroy every nonzero coboundary. Thus we must reduce each component to a single vertex. This can be done by contracting $\alpha_0(G) - p_0(G)$ edges, and no fewer (see 2.213).

CHAPTER 3

Connection

3.1. SEPARATORS

Consider a matroid M on a set E. A *separator* of M is a subset S of E such that each circuit of M is contained either in S or in $E - S$. Any union or intersection of separators of M is a separator, and the complement in E of a separator of M is also a separator.

In particular, E and its null subset are separators of M.

We refer to the minimal non-null separators of M as its *elementary* separators. From the foregoing observations we deduce

3.11. The elementary separators of M are disjoint non-null subsets of E whose union is E.

Separators can be characterized by the following theorem.

3.12. Suppose $S \subseteq E$. Then S is a separator of M if and only if

$$M \cdot S = M \times S.$$

Proof. Suppose S is a separator of M. Then a circuit of M has a non-null intersection with S if and only if it is itself a subset of S. This implies $M \cdot S = M \times S$

Conversely, suppose $M \cdot S = M \times S$. Let Y be any circuit of M. If it meets S it contains a circuit Y_1 of $M \cdot S$, that is, of $M \times S$. But then Y_1 is a circuit of M. Hence $Y = Y_1 \subseteq S$, by Axiom I. We deduce that S is a separator of M.

3.13. Let S be a separator of M. Then, for each $T \subseteq E$, $S \cap T$ is a separator of both $M \cdot T$ and $M \times T$.

27

Proof. Let Y be a circuit of $M \cdot T$ or $M \times T$. Then there is a circuit Z of M such that $Y = Z \cap T$. But either $Z \subseteq S$ or $Z \subseteq E - S$. Hence either $Y \subseteq S \cap T$, or $Y \subseteq T - (S \cap T)$.

3.14. Let S be a separator of M and T a separator of $M \cdot S$, that is, $M \times S$ by **3.12**. Then T is a separator of M.

Proof.

$$M \times T = (M \times S) \times T$$
$$= (M \times S) \cdot T$$
$$= (M \cdot S) \cdot T$$
$$= M \cdot T,$$

by 2.331, 2.334, and 3.12. Hence T is a separator of M by 3.12.

If S is an elementary separator of M, we refer to the matroid $M \cdot S$, that is, $M \times S$, as a *component* of M.

The matroid M is *connected* if it has no separators other than E and its null subset. Thus M is connected if and only if either E is null or M has just one elementary separator.

3.15. If a minor $(M \times S) \cdot T$ of M is connected and T is not null, then T is a subset of an elementary separator of M.

This follows from 3.13.

3.16. Let $M \times S$ and $M \times T$ be connected contractions of M such that $S \cap T$ is non-null. Then $M \times (S \cup T)$ is connected.

Proof. There is an elementary separator Z of $M \times (S \cup T)$ that meets $S \cap T$, by 3.11. It contains both S and T, by 2.331 and 3.15. Hence $Z = S \cup T$, and the theorem follows.

3.2. SEPARABLE AND NONSEPARABLE GRAPHS

Consider a regular chain-group N on a set E. The separators of $M(N)$ are readily interpreted in terms of the properties of N, as the following theorem shows.

3.21. A subset S of E is a separator of $M(N)$ if and only if each chain f of N can be written in the form $g + h$, where g and h are chains of N such that $\|g\| \subseteq S$ and $\|h\| \subseteq E - S$.

Proof. This theorem is a consequence of 1.24.

The theorem can be extended to nonregular chain-groups. A simple induction shows that any nonzero chain of any chain-group N can be expressed as a sum of elementary chains of N, and this result can be used instead of 1.24.

Next we consider a graph G and discuss the separators of $P(G)$ and $B(G)$ in terms of the structure of G.

The graph obtained from G by deleting its isolated vertices is the reduction $G \cdot E(G)$. It evidently has the same polygon-matroid and bond-matroid as G.

If $S \subseteq E(G)$, the common vertices of $G \cdot S$ and $G \cdot [E(G) - S]$ are the *vertices of attachment* of S (and of $E(G) - S$) in G. We denote the set of such vertices by $W(S)$. If $|W(S)| = 1$, then the single vertex of attachment of S is called a *cut-vertex* of G. The graph G is called *separable* if it either has a cut-vertex or is disconnected.

3.22. Let S be a subset of $E(G)$ such that $|W(S)| \leq 1$. Then S is a separator of both $P(G)$ and $B(G)$.

Proof. It is clear that no polygon of G can have edges in both S and $E(G) - S$. Hence S is a separator of $P(G)$.

Let Y be any bond of G with end-graphs H and K. We may suppose K to include no common vertex of $G \cdot S$ and $G \cdot [E(G) - S]$. Then the intersections of K with $G \cdot S$ and $G \cdot [E(G) - S]$ have no common edge or vertex, and since K is connected, one of them is null. Hence either $V(K) \subseteq V(G \cdot S)$ or $V(K) \subseteq V[G \cdot (E(G) - S)]$. Accordingly $Y \subseteq S$ or $Y \subseteq E(G) - S$. The theorem follows.

3.221. If A is a loop of G, then $\{A\}$ is a separator of both $B(G)$ and $P(G)$.

We may note further that $\{A\}$ is a circuit of $P(G)$ and that A belongs to no circuit of $B(G)$.

Let x be any vertex of G. The *star* $\text{St}(x)$ of x in G is the set of all links of G incident with x. We also write $L(x)$ for the set of all loops of G incident with x.

3.23. Let x be a vertex of G such that $\text{St}(x)$ is non-null, and let G_1 be the graph derived from G by deleting the loops of $L(x)$. Then $\text{St}(x)$ is a bond of G if and only if x is not a cut-vertex of G_1.

Proof. Let G_2 be the graph obtained from G by deleting the vertex x and its incident edges. Let those components of G_2 that include vertices incident with members of $\text{St}(x)$ in G be enumerated as $H_1, H_2, \ldots, H_k$. Such components exist since $\text{St}(x)$ is non-null.

Clearly x is a cut-vertex of G_1 if and only if $k \geq 2$.

The set of edges of $\text{St}(x)$ incident in G with vertices of H_i is a bond of G for each i, by Sec. 1.3, one of its end-graphs being H_i. Hence $\text{St}(x)$ is a bond of G if $k = 1$. But if $k \geq 2$, then $\text{St}(x)$ has a bond of G as a proper subset and is not itself a bond of G (see 1.33).

3.24. If $G \cdot E(G)$ is nonseparable, then both $P(G)$ and $B(G)$ are connected matroids.

Proof. Suppose the contrary. Then either $P(G)$ or $B(G)$ has an elementary separator S that is a proper subset of $E(G)$. Using 2.321, 2.323, and 3.12, we deduce from 3.22 that $G \cdot S$ is nonseparable.

Since $\phi \subset S \subset E(G)$, the graph G has at least two edges. Hence it has no loop, by 3.221.

Since $G \cdot E(G)$ is nonseparable S has a vertex x of attachment. Then $\text{St}(x)$ is a bond of $G \cdot E(G)$, by 3.23, and it meets both S and $E(G) - S$. It follows that S is not a separator of $B[G \cdot E(G)]$, that is, $B(G)$.

Let C be the component of $G : [E(G) - S)$ that has x as a vertex. It includes a second vertex y of $G \cdot S$, for otherwise x would be a cut-vertex of $G \cdot E(G)$. By a well-known result in graph theory we can find a simple path P_1 in C from x to y, and we can adjust the notation so that P_1 passes through no vertex of $G \cdot S$ other than x and y. (A "simple" path is a path in which no vertex is repeated.) Similarly there is a simple path P_2 from x to y in the connected graph $G \cdot S$. Combining P_1 and P_2 we derive a polygon of G having edges in both

S and $E(G) - S$. Hence S is not a separator of $P(G) \cdot E(G)$, that is, $P(G)$.

This contradiction establishes the theorem.

3.25. The propositions "$G \cdot E(G)$ is nonseparable," "$P(G)$ is connected," and "$B(G)$ is connected" are equivalent.

We obtain this result by combining 3.22 and 3.24. It shows us that the property of connection in matroids corresponds to that of nonseparability in graphs.

The maximal non-null nonseparable subgraphs of a graph G are called its *blocks* or *cyclic elements*.

If x is an isolated vertex of G, then the *vertex-graph* $[x]$, which has the single vertex x and no edges, is clearly a cyclic element of G. Such cyclic elements we call *degenerate*. The other, nondegenerate, cyclic elements all have edges and they are all without isolated vertices.

3.26. The propositions "$G \cdot S$ is a nondegenerate cyclic element of G," "S is an elementary separator of $P(G)$," and "S is an elementary separator of $B(G)$" are equivalent.

Proof. Let S be any non-null subset of $E(G)$. Suppose $G \cdot S$ is a cyclic element of G. Then by 3.25, with the help of 2.323 and 2.321, we find that $P(G) \times S$ and $B(G) \cdot S$ are connected. Hence, by 3.15 there are elementary separators T_1 and T_2 of $P(G)$ and $B(G)$, respectively, such that $S \subseteq T_1$ and $S \subseteq T_2$.

Conversely, let S be an elementary separator of $P(G)$ or $B(G)$. Then $G \cdot S$ is nonseparable, by 3.25. Hence there is a cyclic element $G \cdot T$ of G such that $S \subseteq T$.

These results establish the theorem.

We must now deal further with the graph-theoretical characterization of cyclic elements.

3.27. Let S be a non-null subset of $E(G)$. Then $G \cdot S$ is a cyclic element of G if and only if the following conditions hold:

(i) $G \cdot S$ is nonseparable.

(ii) Each component of $G : [E(G) - S]$ has at most one vertex in common with $G \cdot S$.

Proof. Suppose first that $G \cdot S$ is a cyclic element of G.

It may happen that S consists of a single loop of G. Then proposition (ii) is trivially true. Hence, by 3.221 we may assume that no edge of S is a loop.

Let x be any vertex of $G \cdot S$ and let T be its star in $G \cdot S$. Then T is a bond of $G \cdot S$ by 3.23. It is thus a circuit of $B(G \cdot S)$, of $B(G) \cdot S$ by 2.321, of $B(G) \times S$ by 3.26 and 3.12, and therefore of $B(G)$. Thus T is a bond of G. Let $C(x)$ be the end-graph of T in G that includes x. Clearly $C(x)$ is a component of $G : [E(G) - S]$, and x is its only common vertex with $G \cdot S$.

Conversely, suppose S to satisfy (i) and (ii). Let C be any component of $G : [E(G) - S]$. Then $E(C)$ has at most one vertex of attachment, by (ii), and therefore any subgraph of G having edges in both S and $E(C)$ is separable. We deduce that S is a maximal nonseparable subgraph of G, that is, a cyclic element of G.

Suppose $G \cdot S$ is a nondegenerate cyclic element of G. For each vertex x of $G \cdot S$ we define $C(G, S, x)$ as that component of $G : [E(G) - S]$ which has x as a vertex. If x and y are distinct vertices of G, then $C(G, S, x)$ and $C(G, S, y)$ can have no common vertex, by 3.27.

3.28. Let G be a connected graph and let S and T be distinct elementary separators of $B(G)$ [or $P(G)$]. Then there are vertices s and t, of $G \cdot S$ and $G \cdot T$, respectively, such that $G \cdot S \subseteq C(G, T, t)$ and $G \cdot T \subseteq C(G, S, s)$. Moreover, each vertex of G belongs to one or both of $C(G, S, s)$ and $C(G, T, t)$.

Proof. The first part of the theorem follows from the fact that $G \cdot S$ is a connected subgraph of $G \cdot [E(G) - T]$, by 3.27, and from the analogous result for $G \cdot T$.

To prove the second part, suppose v is a vertex of G not belonging to $C(G, S, s)$. By the connection of G, v belongs to some $C(G, S, u)$ such that $u \neq s$. But then $C(G, S, u)$, with $G \cdot S$, is a subgraph of $C(G, T, t)$.

3.3. CONNECTED FLATS

Let M be a matroid on a set E. A flat S of M is called "connected" if $M \times S$ is a connected matroid. The lattice of connected flats of M has some interesting properties.

In this discussion we refer to the flats of M of ranks 1, 2, and 3 as the *points, lines,* and *planes* of M, respectively, for the sake of some geometrical analogies. The "points" of M are thus the same as its circuits. We say two flats S and T are *on* one another if either $S \subseteq T$ or $S \supseteq T$. We summarize the properties of lines as follows.

3.31. A disconnected line of M is on just two distinct points, and a connected line is on at least three. If X and Y are distinct points on a line L, then $X \cup Y = L$. Moreover $X \cap Y$ is non-null if and only if L is connected.

Proof. Let L be a line of M. Choose $a \in L$. Then L is on the point $X = \langle L - \{a\} \rangle$ of M, by 2.43. Choose $b \in X$. Then L is also on the point $Y = \langle L - \{b\} \rangle$ of M, and X and Y are distinct.

We have $X \subset X \cup Y \subseteq L$. Hence $r(X \cup Y) = r(L) = 2$ and therefore $X \cup Y = L$, by 2.44.

Suppose L is disconnected. Then X and Y must be separators of L. In this case L can be on no third point of M, by Axiom I. Moreover, X and Y must be disjoint.

Next suppose L is connected. If $X \cap Y$ is null there must be a third point Z of M on L meeting both X and Y. But then $X \cup Z = L$, by one of the preceding results, and therefore $Y \subseteq Z$, which is contrary to Axiom I. We deduce that $X \cap Y$ is non-null. But if $c \in X \cap Y$ there is a third point $\langle L - \{c\} \rangle$ on L, distinct from both X and Y.

We go on to theorems about general connected flats.

3.32. Let S and T be connected flats of M such that $S \subset T$. Then there exists a connected flat U of M, of rank $r(S) + 1$, such that $S \subset U \subseteq T$.

Proof. We may assume S non-null. Since T is connected we can find a point X of M such that $X \subseteq T$, and X meets both S and $T - S$. Choose such an X so that $|X - S|$ has the least possible value.

By 3.16, $S \cup X$ is a connected flat of M. Moreover, $r(S \cup X) > r(S)$, by 2.44.

Suppose $r(S \cup X) > r(S) + 1$. Choose $a \in X - S$. Then

$$r[\langle (S \cup X) - \{a\} \rangle] \geq r(S) + 1,$$

by 2.43. Hence there is a point Y of M such that $Y \subseteq (S \cup X) - \{a\}$, and Y meets $X - S$. But $Y \cap S = \phi$ by the choice of X. Hence $Y \subset X$, contrary to Axiom I. We deduce that in fact $r(S \cup X) = r(S) + 1$. Thus the theorem holds with $U = S \cup X$.

3.33. Let S be a connected flat of M on another connected flat T of M of rank $r(S) + 2$. Then there exist connected flats U and V of M, each of rank $r(S) + 1$, such that $S = \langle U \cap V \rangle$ and $T = U \cup V$.

Proof. We may assume S non-null, by 3.31. As in the preceding proof we can find a point X on T, meeting both S and $T - S$, such that $r(S \cup X) = r(S) + 1$. We write $U = S \cup X$. By repeated application of 2.43 there is a line L on T and X that is on no point of S. Choose $a \in X - S$ and write $Y = \langle L - \{a\} \rangle$.

Suppose Y does not meet $S \cup X$. Since T is connected, it has a point Y_1 that meets both Y and $T - Y$. But Y_1 does not contain Y, by Axiom I. We now have $S \subset U \subset U \cup Y_1 \subset U \cup Y_1 \cup Y \subseteq T$. But then $r(T) \geq r(S) + 3$, by 2.44, which is contrary to hypothesis. We deduce that Y meets either S or X.

Suppose $Y \cap S$ is null. Choose $c \in Y \cap X$ and write $Z = \langle L - \{c\} \rangle$. Then Z meets S, since $Z \cup Y = L$ by 3.31. Since we can replace Y by Z we may assume henceforth that Y meets S. Write $V = S \cup Y$. Then U and V are distinct. Moreover, $S \subset V \subset T$ and therefore $r(V) = r(S) + 1$. The flat V is connected by 3.16.

We have

$$S \subseteq \langle U \cap V \rangle \subset V \subset U \cup V \subseteq T:$$

and therefore $S = \langle U \cap V \rangle$ and $T = U \cup V$, by 2.44.

3.34. Let S, T, and U be flats of M such that S and T are connected, $S \cup U \subseteq T$, and $\langle S \cap U \rangle = \phi$. Then there exists a connected flat R of M such that $S \subseteq R \subseteq T$, $\langle R \cap U \rangle = \phi$, and $r(R) = r(T) - r(U)$.

Proof. If possible choose S, T, and U so that the theorem fails and $r(U)$ has the least rank consistent with this condition. Then $r(U) > 0$, since if $U = \phi$ the theorem holds with $R = T$. Let W be a connected flat of M, of greatest possible rank, such that $S \subseteq W \subset T$ and W does not contain U. We note that $r(W) < r(T)$.

Suppose $r(W) < r(T) - 1$. Then by 3.32 and 3.33 there are distinct connected flats K and L on T, each of rank $r(W) + 1$, such that $\langle K \cap L \rangle = W$. But then K and L cannot both contain U, and this is contrary to the choice of W. We deduce, using 2.44, that $r(W) = r(T) - 1$. We have also by 2.44, $r(\langle U \cap W \rangle) < r(U)$.

By the choice of S, T, and U there is a connected flat R of M such that $S \subseteq R \subseteq W$, $\langle R \cap \langle U \cap W \rangle \rangle = \phi$, and $r(R) = r(W) - r(\langle U \cap W \rangle)$.

We note that any point common to R and U is in W and therefore in $\langle U \cap W \rangle$. We thus have $S \subseteq R \subseteq T$, $\langle R \cap U \rangle = \phi$, and $r(R) \geq r(T) - r(U)$. But we have also $r(T) \geq r(R) + r(U)$, by 2.46. Hence $r(R) = r(T) - r(U)$. Thus the choice of S, T, and U is contradicted. The theorem follows.

3.35. Let L be a disconnected line on a connected flat S of M such that $r(S) \geq 3$. Then there exists a connected plane P of M such that $L \subset P \subseteq S$.

Proof. Let the two points on L be X and Y (see 3.31). Let P be a connected flat of M, of least possible rank, such that $L \subset P \subseteq S$. Assume $r(P) > 3$.

Suppose first there is a second disconnected line L' on X and P. Let its point other than X be Z. Then X, Y, and Z are distinct points of P, by 2.44. By 3.34 there is a connected flat U of M such that $Y \subseteq U \subseteq P$, $r(U) = r(P) - 2$, and $\langle U \cap L' \rangle = \phi$. By 3.33 there are connected flats V and W of M, each of rank $r(P) - 1$, such that $\langle V \cap W \rangle = U$ and $V \cup W = P$. These two flats meet L' in distinct points, by 2.46. Since there are only two points on L' we may suppose $X \subseteq V$. But then L is on V and the definition of P is contradicted.

We deduce that there is no second disconnected line on X and P, and similarly no second disconnected line on Y and P.

Choose $a \in P - L$ and write $R = \langle P - \{a\} \rangle$. Then $L \subseteq R$ and $r(R) = r(P) - 1$, by 2.43. By the definition of P the flat R is disconnected.

Let Z be a separator of R, that is, of $M \times R$, that does not contain X. Let X' be a point of M contained in Z. Then $X \cup X'$ is a flat of M with only two points X and X'. Its rank is thus 2. It is a disconnected line, by 3.31. Hence $X' = Y$. Repeating the argument with Y replacing

X we find that X and Y are the only points of R. Thus $R = L$, $r(R) = 2$, and $r(P) = 3$. Our assumption that $r(P) > 3$ is contradicted. The theorem follows.

3.4. LINEAR SUBCLASSES

Let M be any matroid on a set E, and let C be any class of points of M. We call C a *linear subclass* of M, —strictly, "of the set of points of M"—if it has the following property. If two distinct points X and Y of C are on a common line L, then every point on L is in C.

One example of a linear subclass of M is the set of all points on a given flat.

Another example applies for binary matroids. It depends on the following theorem.

3.41. Let L be a line of a binary matroid M. Then if L is connected there are just three points on L, and each cell of L belongs to just two of them.

Proof. Let X and Y be distinct points on L (see 3.31). We have $M = M(N)$, where N is a binary chain-group. X and Y are domains of chains f and g of N, respectively. The nonzero chain $f + g$ has domain $(X \cup Y) - (X \cap Y)$. Hence there is a point Z of M on L such that Z does not meet $X \cap Y$. Since $Z \cup X = Z \cup Y = L$ we must have $Y - X \subseteq Z$ and $X - Y \subseteq Z$. Hence $Z = (X - Y) \cup (Y - X) = (X \cup Y) - (X \cap Y)$. Hence each cell of L belongs to just two of X, Y, and Z.

Suppose T is a fourth point of L. Since $T \cup X = T \cup Y = T \cup Z = L$ the set T must contain $L - X$, $L - Y$, and $L - Z$. But then $T = L$, contrary to 2.44.

We can now give our second example of a linear subclass. Let us say that a subset S of E *cuts* another subset T of E if both $S \cap T$ and $T - S$ are non-null. Then we have

3.42. Let M be binary. Let C be the class of all points of M that do not cut a given subset S of E. Then C is a linear subclass of M.

Proof. Let X and Y be distinct points of a connected line L of M. Let Z be the third point on L (see 3.41). If $X \cap S$ and $Y \cap S$ are both null, then $Z \cap S$ is null, since $Z \subseteq X \cup Y$. If $S \subseteq X$ and $S \subseteq Y$, then $S \cap Z = \phi$, by 3.41. In each case $Z \in C$.

In the remaining case we may suppose $X \cap S = \phi$ and $S \subseteq Y$. Then $S \subseteq Z$, by 3.41, and so $Z \in C$.

Since a disconnected line L would have only two points, the theorem follows.

We now describe some important properties of linear subclasses of general matroids.

3.43. Let X be a point of a connected flat S of M such that $r(S) \geq 2$. Let C be a linear subclass of M such that not every point on S is in C. Then there exists a point Y on S, distinct from X and not in C, such that $X \cup Y$ is a connected line of M.

Proof. If possible choose X, S, and C so that the theorem fails and $r(S)$ has the least value consistent with this condition. Evidently $r(S) \geq 3$ by the definition of a linear subclass. Using 3.32 we find that there is a connected flat U of M such that $X \subseteq U \subset S$ and $r(U) = r(S) - 2$. By 3.33 there are connected flats V and W of M such that $\langle V \cap W \rangle = U$ and $V \cup W = S$, the flats V and W being each of rank $r(S) - 1$.

By the choice of X, S, and C every point on V or W must belong to C. But there is a point Z on S that is not in C. By 3.34 there is a line L on Z and S that has no common point with U. By 2.46 there are points $Z_1 \subseteq L \cap V$ and $Z_2 \subseteq L \cap W$. These are distinct by the definition of L, and they are also distinct from Z. It follows from the definition of a linear subclass that Z_1 and Z_2 are not both in C. This contradiction establishes the theorem.

3.44. Let C be any linear subclass of M. Let S and T be flats of M, not necessarily connected, such that $S \subseteq T$ and $r(T) = r(S) + 1$. Suppose all the points on S and at least one other point on T belong to C. Then all points on T are in C.

Proof. If the theorem fails we can find points $X \in C$ and $Y \notin C$, each being on T but not S.

Assume $X \cup Y$ is not connected. Then X and Y must be complementary separators of this flat. Accordingly the non-null sets

$X \cap (T - S)$ and $Y \cap (T - S)$ are disjoint. But then $S \subset S \cup X \subset T$ and $r(T) \geq r(S) + 2$, by 2.44. From this contradiction we deduce that $X \cup Y$ is connected.

By 3.43 there is a connected line L of M such that $X \subseteq L \subseteq X \cup Y$ and at least one point of L is not in C. But there is a point Z of M such that $Z \subseteq L \cap S$, by 2.46. By the definition of a linear subclass we must have $Z \notin C$, which is contrary to hypothesis. The theorem follows.

3.45. Let S be a connected flat of M. Let C_1 and C_2 be distinct linear subclasses of M, neither of which includes all the points of S. Then there is a point Z on S that belongs neither to C_1 nor to C_2.

Proof. Choose points X and Y on S such that $X \notin C_1$ and $Y \notin C_2$. We may suppose that X and Y are distinct, that $X \in C_2$, and that $Y \in C_1$, for otherwise the theorem is trivially true.

By 3.43 there is a point W on S, not in C_2, such that $X \cup W$ is a connected line L. We may suppose $W \in C_1$, since otherwise the theorem holds with $Z = W$. But then L has a third point, by 3.31, and this can be in neither C_1 nor C_2, by the definition of a linear subclass.

As important special cases of 3.45 we note the following.

3.46. Let a and b be distinct cells of a connected matroid M. Then there is a point of M that includes both a and b.

3.47. Let M be a binary matroid on a set E. Let S and T be subsets of E, each cut by some circuit of M. Then there exists a circuit of M cutting both S and T.

To prove 3.46 we take C_1 to be the set of points of M on $\langle E - \{a\}\rangle$ and C_2 to be the set on $\langle E - \{b\}\rangle$: The theorem then follows from 3.45. Theorem 3.47 is a consequence of 3.42 and 3.45.

By applying the foregoing theorems to $P(G)$ and $B(G)$ we can obtain theorems about a graph G. The rules of transformation are as follows. A subset S of $E(G)$ is a connected flat of $B(G)$ if and only if $G \times S$ is nonseparable (see 2.322 and 3.25). S is a connected flat of $P(G)$ if and only if $G \cdot S$ is nonseparable (see 2.323 and 3.25). The new theorems thus express properties of the nonseparable contractions and reductions of G.

Theorem 3.46, stated for $P(G)$, tells us that if A and B are distinct edges of a nonseparable graph G, then there is a polygon of G that includes both. This is not essentially different from Whitney's famous result that G has a polygon through any two given vertices.

CHAPTER 4

Bridges

4.1. BRIDGES

In this chapter we take M to be a given matroid on a set E, and we fix some circuit (that is, point) Y of M.

We refer to the elementary separators of $M \cdot (E - Y)$ as the *bridges* of Y in M. Such a bridge is *trivial* if it contains no circuit of $M \cdot (E - Y)$ and *unicellular* if it has just one cell. Every trivial bridge is unicellular, but not every unicellular bridge is trivial.

4.11. $r[M \cdot (E - Y)] = r(M) - 1$.

Proof.

$$r(M) = r(M \times Y) + r[M \cdot (E - Y)],$$
$$= 1 + r[M \cdot (E - Y)]. \qquad \text{by 2.52,}$$

4.12. Suppose $S \subseteq E - Y$ and let k be a non-negative integer. Then S is a flat of $M \cdot (E - Y)$ of rank k if and only if $S \cup Y$ is a flat of M of rank $k + 1$.

Proof. Suppose S is a flat of $M \cdot (E - Y)$. Then each point of $M \cdot (E - Y)$ on S is the intersection with S of a point of M contained in $S \cup Y$. Since S is a union of points of $M \cdot (E - Y)$ it follows that $S \cup Y$ is a union of points of M. It is a "flat" of M.

Conversely, suppose $S \cup Y$ is a flat of M. If $a \in S$ there is a point X of M such that $a \in X \subseteq S \cup Y$. Hence there is a point X' of $M \cdot (E - Y)$ such that $a \in X' \subseteq S$, by 1.11. Hence S is a union of points of $M \cdot (E - Y)$, that is, a flat of $M \cdot (E - Y)$.

41

To complete the proof we observe that

$$r[M \times (S \cup Y)] = r\{[M \times (S \cup Y)] \times Y\}$$
$$+ r\{[M \times (S \cup Y)] \cdot S\},$$

by 2.52,

$$= r(M \times Y) + r\{[M \cdot (E - Y)] \times S\},$$

by 2.331 and 2.334. Here $r[M \times (S \cup Y)]$ is the rank of $S \cup Y$ in M, $r(M \times Y)$ is 1, and $r\{[M \cdot (E - Y)] \times S\}$ is the rank of S in $M \cdot (E - Y)$.

4.13. Let S be a connected flat of $M \cdot (E - Y)$. Then $S \cup Y$ is a connected flat of M unless

$$[M \cdot (E - Y)] \times S = M \times S.$$

If this condition holds, however, then S and Y are the elementary separators of $M \times (S \cup Y)$.

Proof. Let Z be a separator of $M \times (S \cup Y)$. Then either $Y \subseteq Z$ or $Y \cap Z = \phi$. Moreover $Z \cap S$ is a separator of $[M \times (S \cup Y)] \cdot S$, by 3.13, and this matroid is $[M \cdot (E - Y)] \times S$, by 2.334. Hence $S \subseteq Z$ or $S \cap Z = \phi$, by hypothesis. Accordingly, S and Y are the only possible nontrivial separators of $M \times (S \cup Y)$.

The necessary and sufficient condition for S and Y to be separators of $M \times (S \cup Y)$ is

$$[M \times (S \cup Y)] \cdot S = [M \times (S \cup Y)] \times S,$$

by 3.12. This is equivalent to the condition stated in the enunciation, by 2.331 and 2.334.

4.14. Let B be a bridge of Y in M such that $B \cup Y$ is not connected. Then B is an elementary separator of M.

Proof.

$$M \times B = [M \times (B \cup Y)] \times B,$$
$$= [M \times (B \cup Y)] \cdot B, \qquad \text{by 2.331,}$$

by 3.12 and 4.13,

$$= [M \cdot (E - Y)] \times B,$$
$$= [M \cdot (E - Y)] \cdot B, \qquad \text{by 2.334,}$$

by 3.12, since B is a separator of $M \cdot (E - Y)$,

$$= M \cdot B, \qquad \text{by 2.332.}$$

Hence B is a separator of M, by 3.12. It is elementary since $[M \cdot (E - Y)] \cdot B$, that is, $M \cdot B$, is connected (see 3.15).

In what follows we are concerned only with binary matroids and we do not pursue the general theory any further.

4.2. BRIDGES IN BINARY MATROIDS

From now on we suppose M to be binary.

Consider any point Z of $M \cdot \overline{Y}$ (where $\overline{Y} = E - Y$). The flat $Y \cup Z$ of M is a line of M, by 4.12. If it is connected it has just two points other than Z, and their intersections with Y are complementary non-null subsets T and U of Y, by 3.41. These are the *primary segments* of Y determined by Z, and the unordered pair $\{T, U\}$ is the *partition* of Y determined by Z.

If $Y \cup Z$ is not connected, it has only the two points Y and Z, by 3.31. In this case it is convenient to say that the primary segments of Y determined by Z are Y and ϕ, and that the partition of Y determined by Z is the pair $\{Y, \phi\}$.

Let B be any bridge of Y in M. If it is nontrivial we define a *B-segment* of Y as a minimal non-null intersection of primary segments of Y determined by points of $M \cdot \overline{Y}$ on B. If B is trivial we say that the only *B-segment* of Y is Y itself. We thus have

4.21. The B-segments of Y are disjoint non-null subsets of Y whose union is Y.

The class of B-segments of Y is the *partition* of Y determined by B. We denote it by $\pi(M, B, Y)$.

Two bridges B_1 and B_2 of Y in M will be said to *avoid* one another if there exists $S_1 \in \pi(M, B_1, Y)$ and $S_2 \in \pi(M, B_2, Y)$ such that $S_1 \cup S_2 = Y$. If they do not avoid one another they *overlap*.

The point Y of M is *bridge-separable* if its bridges can be classified in two disjoint classes P and Q such that any two members of the same class avoid one another. It is *totally* bridge-separable if no two of its bridges overlap.

We call a binary matroid M an *even* matroid if every point of M is bridge-separable.

4.3. SKEWNESS

By a definition of Sec. 3.4 a point X of M is said to *cut* a subset S of E if neither $X \cap S$ nor $S - X$ is null.

We now say that a point Z of $M \cdot \overline{Y}$ *severs* a subset S of Y if it determines a partition $\{T, U\}$ of Y such that $S \cap T$ and $S \cap U$ are both non-null. Thus Z severs S if and only if the point $Z \cup T$ (or $Z \cup U$) of M cuts S.

4.31. The points of $M \cdot \overline{Y}$ that do not sever a given subset S of Y constitute a linear subclass of $M \cdot \overline{Y}$.

Proof. Let Z_1, Z_2, and Z_3 be the three points of some connected line L of the binary matroid $M \cdot \overline{Y}$ (see 2.36 and 3.41).

Suppose Z_1 and Z_2 do not sever S. Then each point on the lines $Y \cup Z_1$ and $Y \cup Z_2$ of M either contains S or does not meet S. That is, it does not cut S.

Let X_3 be any point of M, other than Y, on the line $Y \cup Z_3$ of M. There is a line L' of M on X_3 and the plane $Y \cup L$ of M (see 4.12) that is not on Y, by 2.43. This line meets $Y \cup Z_1$ and $Y \cup Z_2$ in two distinct points, by 2.46 and 3.31. Since these do not cut S, the point X_3 does not cut it either, by 3.42. It follows that the point Z_3 of $M \cdot \overline{Y}$ does not sever S.

The theorem follows.

We say that two points Z_1 and Z_2 of $M \cdot \overline{Y}$ are *skew* with respect to Y if they determine partitions $\{T_1, U_1\}$ and $\{T_2, U_2\}$ of Y, respectively, such that the sets $T_1 \cap T_2$, $T_1 \cap U_2$, $U_1 \cap T_2$, and $U_1 \cap U_2$ are all non-null.

Extending this definition we say that two bridges B_1 and B_2 of Y in M are *skew* if there are points Z_1 and Z_2 of $M \cdot \overline{Y}$, on B_1 and B_2, respectively, such that Z_1 and Z_2 are skew with respect to Y. We also say that a point Z_1 of $M \cdot \overline{Y}$ is skew to a bridge B_2 of Y in M if there is a point Z_2 of $M \cdot \overline{Y}$ on B_2 that is skew to Z_1.

We proceed to investigate the relation between skewness and overlapping. It is convenient to say that two bridges B_1 and B_2 of Y in M are *equipartite* if

$$\pi(M, B_1, Y) = \pi(M, B_2, Y).$$

We say a bridge B of Y in M is an n-bridge, where n is an integer, if $\pi(M, B, Y)$ has just n members.

4.32. Let B be a 2-bridge or 3-bridge of Y in M, and suppose $W \in \pi(M, B, Y)$. Then there is a point Z of $M \cdot \overline{Y}$ on B that determines the partition $\{W, Y - W\}$ of Y.

Proof. If B is a 2-bridge, this is immediate from the definitions. We may therefore write $\pi(M, B, Y) = \{U, V, W\}$. There is a point of $M \cdot \overline{Y}$ on B that severs $U \cup W$, and another that severs $V \cup W$. By 4.31 and 3.45 there is a point Z of $M \cdot \overline{Y}$ on B that severs both $U \cup W$ and $V \cup W$. This must determine the partition $\{U \cup V, W\}$ of Y.

4.33. Let B_1, B_2, and B_3 be distinct bridges of Y in M such that B_2 is skew both to B_1 and to B_3. Then either there is a point Z_2 of $M \cdot \overline{Y}$ on B_2 that is skew to both B_1 and B_3, or there are points Z_1 and Z_3 of $M \cdot \overline{Y}$ on B_1 and B_3, respectively, such that Z_1, B_2, and Z_3 are mutually skew.

Proof. There are points Z_1, Z_2', Z_2'', and Z_3 of $M \cdot \overline{Y}$ on B_1, B_2, B_2, and B_3, respectively, such that Z_1 is skew to Z_2', and Z_2'' is skew to Z_3.

If Z_1 and Z_3 are skew, the second alternative of the theorem holds.

In the remaining case Z_1 and Z_3 determine partitions $\{S_1, T_1\}$ and $\{S_3, T_3\}$ of Y, respectively, such that $T_1 \cap T_3 = \phi$. Then $T_1 \subseteq S_3$ and $T_3 \subseteq S_1$.

Since Z_2' severs T_1, and Z_2'' severs T_3, there is a point Z_2 of $M \cdot \overline{Y}$ on B_2 severing both T_1 and T_3, by 4.31 and 3.45. This point is skew to Z_1 since it severs S_1 and T_1, and skew to Z_3 since it severs S_3 and T_3. This completes the proof.

4.34. Let B_1 and B_2 be skew bridges of Y in M. Then they overlap.

Proof. There are points Z_1 and Z_2 of $M \cdot \overline{Y}$, on B_1 and B_2, respectively, determining partitions $\{S_1, T_1\}$ and $\{S_2, T_2\}$, respectively, of Y such that $S_1 \cap S_2$, $S_1 \cap T_2$, $T_1 \cap S_2$, and $T_1 \cap T_2$ are all non-null.

Suppose B_1 and B_2 do not overlap. Then we can find $U_1 \in \pi(M, B_1, Y)$ and $U_2 \in \pi(M, B_2, Y)$ such that $U_1 \cup U_2 = Y$. Without loss of generality we may suppose $U_1 \subseteq S_1$ and $U_2 \subseteq S_2$. Then $U_1 \cup U_2$ is disjoint from the non-null subset $T_1 \cap T_2$ of Y, a contradiction.

4.35. Let B_1 and B_2 be overlapping bridges of Y in M. Then either they are skew or they are equipartite 3-bridges.

Proof. Assume B_1 and B_2 are not skew.

Let Z be any point of $M \cdot \overline{Y}$ on B_1 determining a partition $\{S, T\}$ of Y. Suppose S and T each meet more than one member of $\pi(M, B_2, Y)$. Then S and T are each severed by some point of $M \cdot \overline{Y}$ on B_2, and hence there is one such point that severs both, by 4.31 and 3.45. But then B_1 and B_2 are skew, contrary to assumption. We deduce that either S or T is contained in a member of $\pi(M, B_2, Y)$.

Suppose B_1 is a 2-bridge or a 3-bridge. If $W \in \pi(M, B_1, Y)$ we can apply the preceding argument with $W = S$ and $Y - W = T$, by 4.32. Hence there exists $U \in \pi(M, B_2, Y)$ such that either $W \subseteq U$ or $Y - W \subseteq U$. The second alternative must be ruled out; it implies that $W \cup U = Y$ and makes B_1 and B_2 not overlap. We deduce further that B_1 is not a 2-bridge. If it were, we would have $Y - W \in \pi(M, B_1, Y)$, and both of the above alternatives would be ruled out.

Suppose B_1 is a 3-bridge. Write $\pi(M, B_1, Y) = \{U, V, W\}$. There are members U', V', and W' of $\pi(M, B_2, Y)$ such that $U \subseteq U'$, $V \subseteq V'$, and $W \subseteq W'$. These three B_2-segments are distinct. If for example U' and V' were the same we would have $W \cup U' = Y$, contrary to the hypothesis that B_1 and B_2 overlap. It follows that B_1 and B_2 are equipartite 3-bridges.

In the remaining case B_1 is an n-bridge, with $n \geq 4$.

Let W_1, W_2, W_3, W_4 be distinct members of $\pi(M, B_1, Y)$. Using 4.31 and 3.45 we find that there is a point Z of $M \cdot \overline{Y}$ on B_1 that cuts both $W_1 \cup W_2$ and $W_3 \cup W_4$. It determines a partition $\{S, T\}$ of Y such that W_1 and W_2 are contained in different sets S and T, as are also W_3 and W_4.

By the first part of the proof we can find $U' \in \pi(M, B_2, Y)$ such that U' contains two of the four given B_1-segments. Now let W' be a member of $\pi(M, B_2, Y)$ containing the greatest possible number k of B_1-segments. The preceding result shows that $k \geq 2$.

Suppose there are two distinct B_1-segments not in W'. Then by our result for four B_1-segments W_1, W_2, W_3, and W_4 it follows that some point of $M \cdot \overline{Y}$ on B_1 determines a partition $\{S, T\}$ of Y such that S and T each meet both W' and $Y - W'$. But some member of $\pi(M, B_2, Y)$ must contain S or T. Since W' meets both without containing either, we have a contradiction.

We deduce that there is at most one B_1-segment that is not contained in W'. This contradicts the hypothesis that B_1 and B_2 overlap.

For completeness we note the obvious fact that two equipartite 3-bridges of Y in M must overlap.

4.4. BRIDGES IN POLYGON-MATROIDS

Let G be a graph and let $G \cdot Y$ be one of its polygons. Write $H = G \times [E(G) - Y]$. Thus H is obtained from G by contracting the polygon $G \cdot Y$ to a single vertex.

We have

$$P(H) = P(G) \cdot [E(G) - Y],$$

by 2.324. We deduce that the bridges of Y in $P(G)$ are the elementary separators of $P(H)$. Thus a subset B of $E(G)$ is a bridge of Y in $P(G)$ if and only if $H \cdot B$ is a cyclic element of H, by 3.26.

Now let B be such a bridge and consider the graph $G \cdot (B \cup Y)$. If this is separable it has just two cyclic elements, defined by B and Y, and moreover $G \cdot B$ is a cyclic element of G, by 4.13 and 4.14. Then $G \cdot B$ has at most one common vertex with the polygon $G \cdot Y$, by

3.27. If $G \cdot (B \cup Y)$ is nonseparable, then $G \cdot B$ and $G \cdot Y$ have $k \geq 2$ common vertices. These subdivide the polygon $G \cdot Y$ into k edge-disjoint "residual arcs of B in $G \cdot Y$."

Consider any polygon Z of $H \cdot B$. We know from 4.12 that $Y \cup Z$ is a line of $P(H)$. Let us digress to interpret the lines of a polygon-matroid in terms of graph structure. We define a θ-*graph* as a graph that is the union of three arcs with the same ends x and y, provided that no two of the arcs have any edge or internal vertex in common. We call x and y the *ends* of the θ-graph.

4.41. Let G be a graph and let L be a subset of $E(G)$. Then L is a disconnected line of $P(G)$ if and only if $G \cdot L$ is the union of two polygons with at most one vertex in common. It is a connected line of G if and only if $G \cdot L$ is a θ-graph.

Proof. Suppose $G \cdot L$ is either (i) a union of two polygons with at most one common vertex or (ii) a θ-graph. Then L is a flat of $P(G)$. We can destroy all the polygons in $G \cdot L$ by deleting two, but not fewer, edges. Hence $r(L) = 2$ in $P(G)$, and L is a line of $P(G)$. It is clearly disconnected in case (i) and connected in case (ii) (see 3.31).

Conversely, suppose L is a line of $P(G)$. Let X and Y be two circuits (points) of $P(G)$ on L. If L is disconnected they are elementary separators of $P(G) \times L$. Hence $G \cdot L$ is the union of two polygons $G \cdot X$ and $G \cdot Y$ with at most one common vertex, by 2.323 and 3.27. If L is connected, $X \cap Y$ is non-null. Hence two distinct vertices x and y of the polygon $G \cdot X$ are joined by an arc in $G \cdot Y$ that has no edge or internal vertex in common with $G \cdot X$. Adjoining this arc to $G \cdot X$ we obtain a θ-graph $G \cdot L'$. Since L' is a line of $P(G)$ by the result already proved, we have $L' = L$, by 2.44.

We return to the polygon Z of $H \cdot B$. Now $G \cdot Z$ may be a polygon of G with at most one vertex in common with $G \cdot Y$, or it may be an arc in G with its two ends x and y, but no other edge or vertex, in $G \cdot Y$. In the first case, $Y \cup Z$ is a disconnected line of $P(G)$, and Z determines the trivial partition $\{Y, \phi\}$ of Y. In the second case, $Y \cup Z$ is a connected line determining a partition $\{T, U\}$ of Y, where $G \cdot T$ and $G \cdot U$ are the two arcs in $G \cdot Y$ joining x and y.

It can be shown by the methods of graph theory that if p and q are distinct common vertices of $G \cdot Y$ and $G \cdot B$ they can be joined by an arc $G \cdot Z$ in $G \cdot B$ with no other vertex in $G \cdot Y$. Then of course $H \cdot Z$ is a polygon.

We conclude from the above observations that the members of $\pi[P(G), B, Y]$ are the edge-sets of the residual arcs of B in $G \cdot Y$. Moreover, each one of them is a primary segment of Y determined by some circuit Z of $P(G) \cdot B$. The latter property of polygon-matroids does not extend to all binary matroids.

The above considerations are helpful in the study of planar graphs. Suppose G is drawn in the 2-sphere so that $G \cdot Y$ is represented by a simple closed curve with residual domains D and D'. It is not difficult to show that if B is a bridge of Y in $P(G)$, then $G \cdot B$ must be drawn entirely in one of D and D', except for its common vertices with $G \cdot Y$. It is also found that two overlapping bridges cannot be represented in the same residual domain. The conclusion is that if G is planar, its polygon-matroid is even. There is a more difficult converse result that if $P(G)$ is even, then G is planar, and a proof of Kuratowski's Theorem can be based on this.

4.5. BRIDGES IN BOND-MATROIDS

Let G be any graph and let Y be one of its bonds. Write $H = G \cdot [E(G) - Y]$. Then H is obtained from G by deleting the edges of Y. We have

$$B(H) = B(G) \cdot [E(G) - Y],$$

by 2.321. We deduce that the bridges of Y in $B(G)$ are the elementary separators of $B(H)$. Thus a subset B of $E(G)$ is a bridge of Y in $B(G)$ if and only if $H \cdot B$ is a cyclic element of H, by 3.26.

Now let B be such a bridge. We note that $H \cdot B$, being a nonseparable graph, is contained in some component C of H, and is a cyclic element of that component, by 3.27.

Among the components of H we have the two end-graphs of Y in B. If there is not more than one bridge of Y in $B(G)$ one of these components is edgeless. Thus we have

4.51. If $B(G) \cdot [E(G) - Y]$ is a connected matroid, then Y is the star of some vertex v in G.

This result is often helpful in determining whether a given binary matroid can be represented as the bond-matroid of a graph.

We proceed to characterize the lines of $B(G)$ in terms of the structure of G.

4.52. Let G be a connected graph and let L be a subset of $E(G)$. Then L is a line of $B(G)$ if and only if there are three distinct components C_1, C_2, C_3, of $G : [E(G) - L]$ with the following properties:

    (i) Each end in G of an edge of L belongs to one of the $V(C_i)$
       ($i = 1, 2, 3$).
   (ii) No edge of L has both ends in the same set $V(C_i)$.
  (iii) Each of the three sets $V(C_i)$ contains an end of an edge of L.

Proof. Suppose first that L satisfies conditions (i), (ii), and (iii). We define L_{ij}, where $1 \leq i < j \leq 3$, as the set of all edges of L with one end in $V(C_i)$ and the other in $V(C_j)$. We can clearly adjust the notation so that L_{12} and L_{23} are non-null.

Consider any bond X of G such that $X \subseteq L$. Any end-graph of X in G must either contain a given C_j or be disjoint from it. The end-graph must also of course be connected.

If $L_{13} = \phi$, it follows that the only bonds of G contained in L are L_{12} and L_{23}. Accordingly L is then a disconnected line of $B(G)$. But if L_{13} is non-null, L contains just three bonds of G, namely, $L_{12} \cup L_{13}, L_{12} \cup L_{23}$, and $L_{13} \cup L_{23}$. In this case L is evidently a connected line of $B(G)$.

Conversely, let L be a line of $B(G)$. We can find two distinct bonds X and Y of G such that $X \cup Y = L$, by 3.31.

Let H and K be the end-graphs of X in G. Without loss of generality we can suppose $Y \cap E(H)$ to be non-null. This set contains a circuit Y_1 of $B(G) \cdot E(H) = B[G \cdot E(H)] = B(H)$, by 2.321. Thus Y_1 is a bond of H. Let its end-graphs in H be H_1 and H_2.

Considering the three components H_1, H_2, and K of $G : [E(G) - (X \cup Y_1)]$ and using the result already proved, we find that $X \cup Y_1$ is a line of $B(G)$. Hence $X \cup Y_1 = L$, by 2.44, and H_1, H_2, and K may be taken as C_1, C_2, and C_3.

We have incidentally proved the following.

Corollary. Suppose L satisfies (i), (ii), and (iii), and let L_{ij} be defined as in the above proof. Then L is a connected line of $B(G)$ if the sets L_{12}, L_{23}, and L_{13} are all non-null, and a disconnected line otherwise.

We return to the consideration of a bridge B in $B(G)$ corresponding to a bond Y of G.

Consider any bond Z of $H \cdot B$ and let its end-graphs in $H \cdot B$ be K_1 and K_2. We know from 4.12 that $Y \cup Z$ is a line of $B(G)$.

Now if $H \cdot B$ is not contained in an end-graph of Y it is clear that the line $Y \cup Z$ is disconnected. Then Z determines the trivial partition $\{Y, \phi\}$ of Y. We suppose therefore that $H \cdot B(= G \cdot B)$ is contained in an end-graph J_1 of Y in G, and we denote the other end-graph of Y by J_2.

Each component of $J_1 : [E(J_1) - B]$ includes exactly one vertex of $G \cdot B$, by 3.27 and the fact that J_1 is connected. If $x \in V(G \cdot B)$, then in accordance with the notation of Sec. 3.2 we denote the component containing x by $C(J_1, B, x)$.

We write $Y(B, x)$ for the set of all edges of Y having one end in $C(J_1, B, x)$. The sets $Y(B, x)$, taken for all $x \in V(G \cdot B)$ are disjoint and have Y as their union.

Write

$$U_i = \bigcup_{x \in V(K_i)} Y(B, x), \qquad (i = 1, 2).$$

If one of the sets U_i is null, then $Y \cup Z$ is a disconnected line by the corollary to 4.52. The graph C_i, where $i = 1$ or 2, may be taken to be the union of K_i and the graphs $C(J_1, B, x)$ with $x \in V(K_i)$. C_3 may be identified with J_2. In this case Z determines the trivial partition $\{Y, \phi\}$ of Y.

But if U_1 and U_2 are both non-null, $Y \cup Z$ is a connected line by the above corollary. Moreover, $Z \cup U_1$ and $Z \cup U_2$ are then bonds of G. We deduce that Z determines the partition $\{U_1, U_2\}$ of Y. If $x \in V(G \cdot B)$ we can take Z to be the star of x in $G \cdot B$, by 3.32. Then we can take K_1 to be the vertex-graph $[x]$. We then find that $U_1 = Y(B, x)$.

From these results we conclude that if $G \cdot B$ is contained in an end-graph of Y in G, then the B-segments of Y in $B(G)$ are the non-null sets $Y(B, x)$, $x \in V(G \cdot B)$. Moreover, each B-segment occurs as a primary segment of Y determined by some circuit Z of $B(G) \cdot [E(G) - Y]$.

The property that each B-segment of Y occurs as a primary segment has now been shown to hold for bond-matroids as well as for polygon-matroids. In Chapter 5 it is extended to all regular matroids.

We call a matroid *graphic* or *cographic* if it can be represented as the bond-matroid or polygon-matroid, respectively, of some graph.

We can use the result so far obtained in this section to establish the following theorem.

4.53. Every graphic matroid is even.

Proof. Let M be a graphic matroid. There is a graph G such that $M = B(G)$. Let Y be any circuit of M, that is, any bond of G, and let the end-graphs of Y in G be J_1 and J_2.

Let B_1 and B_2 be overlapping bridges of Y in $B(G)$.

We have seen that if $G \cdot B_1$ is not contained in an end-graph of Y, then each of its bonds determines a trivial partition of Y. We thus have $\pi(M, B_1, Y) = \{Y\}$ so that B_1 does not overlap B_2. From this contradiction we deduce that $G \cdot B_1$, and similarly $G \cdot B_2$, is contained in an end-graph of Y.

Assume that $G \cdot B_1$ and $G \cdot B_2$ are contained in the same end-graph of Y, which we may suppose to be J_1. There are vertices x_1 of $G \cdot B_1$ and x_2 of $G \cdot B_2$ such that $G \cdot B_2$ is a subgraph of $C(J_1, B_1, x_1)$ and $G \cdot B_1$ is a subgraph of $C(J_1, B_2, x_2)$, by 3.28. Moreover,

$$Y(B_1, x_1) \cup Y(B_2, x_2) = Y,$$

by the same theorem. But $Y(B_1, x_1)$ is either null or a B_1-segment of Y in $B(G)$, and similarly for $Y(B_2, x_2)$. This contradicts the hypothesis that B_1 and B_2 overlap.

We deduce that if B_1 and B_2 overlap, then $G \cdot B_1$ and $G \cdot B_2$ are subgraphs of different end-graphs of Y in G.

Let P_1 be the class of all bridges B of Y in M such that $G \cdot B \subseteq J_1$, and let P_2 be the class of all other bridges of Y in M. Then by the result just proved no two members of the same class P_1 or P_2 overlap. Thus Y is bridge-separable. The theorem follows.

4.6. BRIDGES OF COLLINEAR CIRCUITS

Let M be a binary matroid, let Y be one of its circuits, and let B be a bridge of Y in M. Let Z be a point of $M \cdot \overline{Y}$ on B determining a partition $\{S, T\}$ of Y, where S and T are both non-null.

We note that $Y \cup Z$ is a connected line of M and that the three points on it are Y, $S \cup Z$, and $T \cup Z$. Write $S \cup Z = Y'$.

In this section we discuss the relation between the bridges of Y and those of Y' in M.

4.61. *T is a circuit of* $M \cdot (E - Y')$.

Proof. Since $Z \cup T$ is a circuit of M there is a circuit T_1 of $M \cdot (E - Y')$ such that $T_1 \subseteq (Z \cup T) \cap (E - Y') = T$. There is a circuit X of M such that $X \cap (E - Y') = T_1$. But then X is a point of M on the line $Y' \cup T = Y \cup Z$. Since the only points on this line are Y, $S \cup Z$, and $T \cup Z$, we must have $X = Y$ or $X = T \cup Z$. In either case, $T_1 = T$.

Corollary. There is a bridge B_T of Y' in M such that $T \subseteq B_T$.

4.62. Let C be any bridge of Y in M other than B. Then there is a bridge C' of Y' in M such that $C \subseteq C'$.

Proof. We have $C \subseteq E - Y'$, and

$$[M \cdot (E - Y')] \cdot C = [M \cdot (E - Y)] \cdot C,$$

by 2.332. But the expression on the right represents a connected matroid. Hence C is contained in some elementary separator C' of $M \cdot (E - Y')$, by 3.15.

4.63. Let C be any bridge of Y in M other than B. Suppose T is a subset of some $W \in \pi(M, C, Y)$. Then $C = C'$.

Proof. Let Z_1 be any circuit of $M \cdot (E - Y')$ on C' that meets C.

There is a circuit X of $M \times (C' \cup Y')$ such that $X \cap C' = Z_1$. ($X \cap Y'$ is one of the primary segments of Y' determined by Z_1.) Now X is a circuit of M meeting C. Hence there is a circuit Z_2 of $M \cdot (E - Y)$ such that $Z_2 \cap C$ is non-null and $Z_2 \subseteq X \cap (E - Y)$, by 1.11. Then

$$Z_2 \subseteq Z_2 \cap C \subseteq Z_2 \cap C' \subseteq X \cap C' = Z_1,$$

since C is a separator of $M \cdot (E - Y)$.

Z_2 determines a partition $\{S_2, T_2\}$ of Y, and by our hypothesis we may suppose $T \subseteq W \subseteq T_2$.

The circuit $Z_2 \cup S_2$ of M meets C and therefore C'. Hence there is a circuit Z_3 of $M \cdot (E - Y')$ on C' such that $Z_3 \subseteq (Z_2 \cup S_2) \cap (E - Y')$. But S_2 does not meet $E - Y'$, since $(E - Y') \cap Y = T \subseteq T_2$. Hence $Z_3 \subseteq Z_2$.

We now have $Z_3 \subseteq Z_2 \subseteq Z_1$. Applying Axiom I to $M \cdot (E - Y')$ we find that $Z_1 = Z_2 = Z_3$. Hence $Z_1 \subseteq C$, since Z_2 meets C.

We deduce that C is a separator of $[M \cdot (E - Y')] \times C' = M \cdot C'$, by 3.12 and 2.332. Hence $C = C'$ by the connection of $M \cdot C'$.

4.64. Under the conditions of 4.63, $\pi(M, C, Y')$ differs from $\pi(M, C, Y)$ only in the replacement of W by $(W - T) \cup Z$.

Proof. We note first that

$$[M \cdot (E - Y')] \times C = [M \cdot (E - Y')] \cdot C$$
$$= [M \cdot (E - Y)] \cdot C$$
$$= [M \cdot (E - Y)] \times C,$$

by 3.12 and 2.332. Hence $M \cdot (E - Y)$ and $M \cdot (E - Y')$ have the same circuits on C.

If one such circuit Z_0 determines a partition $\{S_0, T_0\}$ of Y we may suppose $T \subseteq W \subseteq T_0$. But then $S_0 \cup Z_0$ is one of the points on the line $Y' \cup Z_0$. That is, Z_0 determines the partition $\{S_0, (T_0 - T) \cup Z\}$ of Y'. The theorem now follows from the definition of a C-segment.

4.65. Let C be any bridge of Y in M other than B. Suppose T is not contained in any member of $\pi(M, C, Y)$. Then $C' = B_T$.

Proof. There is a circuit X of $M \times (C \cup Y)$ that cuts T. Hence there is a circuit Z_1 of $M \cdot (E - Y')$ that meets T and satisfies $Z_1 \subseteq X \cap (E - Y') \subseteq (C \cup Y) \cap (E - Y') \subseteq C' \cup T$. We note also that Z_1 cuts T.

Now T is another circuit of $M \cdot (E - Y')$, by 4.61. Hence Z_1 is not a proper subset of T, by Axiom I. Accordingly, Z_1 meets both C' and T. Since C' is a separator of $M \cdot (E - Y')$ it follows first that $Z_1 \subseteq C'$ and then that $T \subseteq C'$. Hence $C' = B_T$.

4.66. Suppose T is contained in some member of $\pi(M, C, Y)$, for each bridge C, other than B, of Y in M. Then $B_T - B = T$ (by 4.63).

4.67. Let the bridges of Y in M contained in B_T be enumerated as $B_1, B_2, \ldots, B_k$, where we suppose $k \geq 1$. Let P be the collection of all non-null sets of the form

$$S \cap \bigcap_{i=1}^{k} W_i, \qquad W_i \in \pi(M, B_i, Y).$$

Then the members of P are those members of $\pi(M, B_T, Y')$ that do not meet Z.

Proof. Choose $R \in P$ and suppose R is not contained in any member of $\pi(M, B_T, Y')$. Then there is a point Z_1 of $M \cdot (E - Y')$ on B_T that severs R, and a corresponding point Z_2 of M on the line $Y' \cup Z_1$ that cuts R.

The points of $M \times (B_T \cup Y \cup Z)$ that do not cut R constitute a linear subclass of that matroid, by 3.42. The subclass contains Y but not Z_2. Hence, by 3.43, there is a point Z_3 of $M \times (B_T \cup Y \cup Z)$, collinear with Y, that cuts R.

Now $Z_3 - Y$ is a point of $M \cdot (E - Y)$, by 4.12, and it severs R. It is therefore not Z. It must therefore belong to some bridge B_i of Y in M, with $1 \leq i \leq k$. But this is impossible since the points of B_i do not sever R. We deduce that there is a member R' of $\pi(M, B_T, Y')$ such that $R \subseteq R'$. Since T determines the partition $\{S, Z\}$ of Y', we have $R' \cap Z = \phi$.

Next we suppose that for some B_i, with $1 \leq i \leq k$, there is a member W_i of $\pi(M, B_i, Y)$ that cuts R'. Applying 3.42 and 3.43 as before, we find that there is a member of $\pi(M, B_T, Y')$ that cuts R'. This being impossible, we deduce that $R' \subseteq R$. The theorem follows.

CHAPTER 5

Regular Chain-Groups and Matroids

5.1. REPRESENTATIVE MATRICES

Let N be a chain-group on a set E over a commutative ring R with a unit element and no divisors of zero. We may define the *rank* $r(N)$ of N as the rank of $M(N)$ (see Sec. 2.6).

Consider a set $B = \{f_1, f_2, \ldots, f_r\}$ of $r = r(N)$ linearly independent chains of N. Then every other chain of N is linearly dependent on B, that is, for each $f \in N$ there exists $\lambda \in R$ such that $\lambda \neq 0$ and λf is a sum of multiples of elements of B by members of R. If we can take $\lambda = 1$ for each f we call B a *basis* of N. This is always possible when R is a field, since we can then multiply the expression for λf by the reciprocal of λ.

We denote the number of elements of E by n. It is convenient to represent the set B by a matrix $K = \{k_{ij}\}$ of r rows and n columns, where the elements of E are enumerated as $e_1, e_2, \ldots, e_n$, and k_{ij} is the coefficient of e_j in f_i. If B is a basis of N we call K a *representative matrix* of N. If we know only that the r members of B are linearly independent we call B a *weakly representative matrix* of N.

For any chain f on E over R we call the 1-rowed matrix

$$\{f(e_1), f(e_2), \ldots, f(e_n)\}$$

the *representative vector* of f. Thus the rows of K are the representative vectors of the chains f_i. For most purposes it is immaterial whether we deal with the chains of N or with their representative vectors; any linear combination of representative vectors is the representative vector of the corresponding linear combination of chains.

57

By standard results of linear algebra the property of being a representative matrix of N is invariant under the following "elementary operations":

(i) Permuting the rows.

(ii) Adding to one row a multiple of another by an element of R.

(iii) Multiplying a row by -1.

The property of being a weakly representative matrix of N is invariant under the operation of multiplying a row by any nonzero element of R.

Let K be a weakly representative matrix of N and let S be any subset of E. We define $K(S)$ as the submatrix of K consisting of those columns that correspond to members of S. If S is null we say that K is a "null" matrix having no columns but r rows. A similar convention can be made to justify the above theory when $r(N) = 0$. Then the only basis of N is the null set of chains, and K must be a "null" matrix of no rows but n columns.

If $|S| = r$ it may happen that $K(S)$ is a *diagonal matrix*, that is, the diagonal elements are all nonzero and the nondiagonal elements are all zero. We then say that K is in *diagonal form* with respect to S. We make use of the following rule.

5.11. Suppose K is a weakly representative matrix of N, and that S is a subset of E of r elements. Then we can find a weakly representative matrix J of N that is in diagonal form with respect to S if and only if $K(S)$ is nonsingular.

Proof.   Suppose $K(S)$ is nonsingular, that is, its rows are linearly independent so that its determinant is nonzero. Form a matrix J by premultiplying K by the adjugate matrix of $K(S)$. Then the part of J corresponding to S is diagonal, and hence the rows of J are linearly independent. But these rows are linear combinations, with multipliers in R, of rows of K. Hence J is the required weakly representative matrix.

Next suppose $K(S)$ is singular. Then there is a nonzero linear combination W of rows of K in which each member of S has a zero coefficient. W cannot be a linear combination of rows of a weakly representative matrix J of N that is in diagonal form with respect to S. Hence no such matrix J can exist.

Weakly representative matrices in diagonal form can be used to obtain information about the structure of $M(N)$, by means of the following theorems.

5.12. Let K be a weakly representative matrix of N, diagonal with respect to some $S \subseteq E$. Let s_i be the ith element of S and let f_i be the chain of N represented by the ith row of K. Then f_i is elementary. Moreover, if f is any chain of N such that $\|f\| \cap S = \{s_i\}$, then $\|f\| = \|f_i\|$.

Proof. If the theorem fails we can find a chain f of N such that $\|f\| \neq \|f_i\|$ and $\|f\| \cap S = \|f_i\| \cap S = \{s_i\}$ (by 1.11 if f_i is not elementary). Write

$$h = f_i(s_i)f - f(s_i)f_i .$$

Then $\|h\| \cap S = \phi$. Hence $h = 0$, for otherwise it could not be linearly dependent on the chains represented by the rows of K. But then $\|f\| = \|f_i\|$, contrary to the definition of f.

5.13. Let f be any elementary chain of N. Then we can find a weakly representative matrix K of N, diagonal with respect to some $S \subseteq E$ such that some row of K is the representative vector of f.

Proof. We first investigate the chain-group $N_1 = N \cdot (E - \|f\|)$, whose rank is $r(N) - 1$, by 2.31 and 2.52. We find a weakly representative matrix J of N_1 and a square submatrix $J(S_1)$ of order $r(N_1)$ that is nonsingular. We then reduce J to diagonal form with respect to S_1 (see 5.11). J is thus transformed into a matrix K_1.

Choose $a \in \|f\|$ and write $S = S_1 \cup \{a\}$. Then $|S| = r(N)$. Let f_i, where i ranges from 2 to $r = r(N)$, be a chain of N whose restriction to $E - \|f\|$ is represented by the $(i - 1)^{th}$ row of K_1. Write

$$g_1 = f ,$$
$$g_i = f(a)f_i - f_i(a)f, \quad (2 \leq i \leq r) .$$

Then the r chains g_j of N are linearly independent, defining a weakly representative matrix K of N that is diagonal with respect to S.

Under the conditions of 5.13 it is clear that one row of $K(E - \|f\|)$, corresponding to f, is zero and that the other rows constitute a weakly representative matrix of $N \cdot (E - \|f\|)$, diagonal with respect to $S - \|f\|$.

Let K be a weakly representative matrix of N diagonal with respect to a subset S of E. It may happen that $K(S)$ is the unit matrix. In this case K is a true representative matrix of N since an arbitrary chain of N can be written as

$$f = \sum_{i=1}^{r} f(s_i) f_i \, ,$$

where s_i is the ith element of S, and f_i is the chain represented by the ith row of K. We then call K a *standard* representative matrix of N associated with S.

If R is a field, K can always be reduced to standard form with respect to S by multiplying the rows by appropriate elements of R. A similar reduction can be made if N is regular. For then each row of K represents an integer multiple of a primitive chain, by 5.12, and we need only replace each row by the representative vector of the corresponding primitive chain. This is done by dividing the row by its maximum element.

Combining this last observation with 5.11 we have

5.14. Let K be any weakly representative matrix of a regular chain-group N on E. Let S be a subset of E of r elements, where $r = r(N)$. Then we can find a standard representative matrix of N associated with S if and only if $K(S)$ is nonsingular.

5.2. REPRESENTATIVE MATRICES OF REGULAR CHAIN-GROUPS

In this section we characterize regularity, for an integral chain-group N, in terms of the representative matrices of N.

5.21. Let N be a regular chain-group on a set E. Let K be a standard representative matrix of N associated with a subset S of E. Let T be any subset of E such that $|T| = |S| = r(N)$. Then the determinant of $K(T)$ has one of the values 1, –1, or 0.

Proof. Suppose det $K(T) \neq 0$. There is a standard representative matrix K_1 of N associated with T, by 5.14. Since the rows of K_1 are linear combinations, with integer multipliers, of those of K there is a square matrix A of integers such that $AK = K_1$. But then

$$AK(T) = K_1(T) .$$

Taking determinants we have

$$\det A \cdot \det K(T) = 1,$$

since $K_1(T)$ is a unit matrix. Since A and $K(T)$ are matrices of integers it follows that $\det K(T) = \pm 1$.

Corollary. The determinants of the square submatrices of $K(E - S)$ are restricted to the values 1, –1, and 0.

Proof. Let A be such a submatrix. Let T be the subset of E corresponding to the columns of $K(E - S)$ meeting A and those of $K(S)$ with no 1's in rows meeting A. Then $K(T)$ is square, and $\det K(T) = 1, -1$, or 0, by 5.21. But expansion of $\det K(T)$, using the columns in $K(S)$, shows that $\det K(T) = \pm \det A$.

We express the result of the corollary by saying that $K(E - S)$ is *completely unimodular*. Evidently K itself is completely unimodular.

5.22. Let K be a weakly representative matrix of an integral chain-group N, of rank r. Suppose each $r \times r$ submatrix of K has a determinant equal to 1, –1, or 0. Then N is regular and K is a true representative matrix of N.

Proof. Let f be any elementary chain of N. By 5.13 there is a weakly representative matrix J of N, diagonal with respect to some $S \subseteq E$ such that some row of J represents f.

We have $\det K(S) = \pm 1$, by 5.11 and our hypothesis. Hence $[K(S)]^{-1}$ is a matrix of integers. Multiplying K, in front, by $[K(S)]^{-1}$ we obtain a standard representative matrix K_1 of N associated with S, still with the property that $\det K_1(T) = 1, -1$, or 0 if $|T| = r$.

One row of K_1, say the ith, represents a chain g of N such that $\|g\| = \|f\|$, by 5.12. Consider any $b \in \|g\| - S$. Let S' be derived from S by deleting the cell of $\|g\| \cap S$ and adjoining b. Then $|S'| = r$, and so det $K(S') = 1$, -1, or 0. But det $K(S') = \pm g(b)$, by expansion in terms of the columns in $K(S)$.

We deduce that the coefficients of g are restricted to the values 1, -1, and 0. Hence g is a primitive chain and f must be an integral multiple [by $f(b)g(b)$] of g. It follows that N is regular.

Now K_1 is a standard representative matrix of N. Hence each representative vector of a chain of N is a linear combination with integer multipliers, of rows of K_1, and therefore of rows of K. Accordingly, K is a representative matrix of N.

5.23. Let K be a standard representative matrix, associated with some $S \subseteq E$, of an integral chain-group N on E. Suppose $K(E - S)$ is completely unimodular. Then N is regular.

Proof. Let T be any subset of E having just $r = r(N)$ elements. If $T = S$ we have det $K(T) = 1$. If $T \neq S$ let A be the square submatrix of $K(E - S)$ defined by the intersection of the columns in $K(T)$ with the rows of K having no 1's in $K(T \cap S)$. Then det $K(T) = \pm \det A$, by expansion in terms of the columns of $K(T)$ in $K(S)$.

In every case we have det $K(T) = 1$, -1, or 0, by our hypothesis. Hence N is regular, by 5.22.

Consider a standard representative matrix K, with respect to some $S \subseteq E$, of a regular chain-group N on E. We suppose $r(N) = r$ and $|E| = n$. Let us construct an $(n - r) \times n$ matrix K^* as follows. The columns corresponding to those of $K(E - S)$ are occupied by a unit matrix, and the complementary submatrix to this, consisting of the remaining columns of K^*, is minus the transpose of $K(E - S)$.

It is easily verified that each row of K^* is orthogonal to each row of K, and therefore represents a chain of N^*. On the other hand, let f be any chain of N^*. By adding integral multiples of the chains of N^* represented by rows of K^* we can obtain a chain g of N^* such that $\|g\| \subseteq S$. But then $g = 0$, since it is orthogonal to the chains represented by rows of N.

We deduce that K^* is a standard representative matrix of N^* associated with $E - S$. Using 5.23 we obtain

5.24. If N is a regular chain-group on E, then N^* is regular and $r(N) + r(N^*) = |E|$.

If we repeat the above construction starting with K^* we recover K. We deduce

5.25. If N is a regular chain-group, then $(N^*)^* = N$.

5.26. Let N be a regular chain-group on E and let S be any subset of E. Then

$$(N \cdot S)^* = N^* \times S,$$

$$(N \times S)^* = N^* \cdot S.$$

Proof. Let f be any integral chain on S. It belongs to $(N \cdot S)^*$ if and only if it is orthogonal to every chain of $N \cdot S$. Let f' be the chain on E whose restriction to S is f and whose restriction to $E - S$ is zero. Then $f \in (N \cdot S)^*$ if and only if $f' \in N^*$, that is, if and only if $f \in N^* \times S$.

This establishes the first identity. Applying it to the regular chain-group N^* we have, after taking duals,

$$[(N^* \cdot S)^*]^* = [(N^*)^* \times S]^*.$$

But $(N^*)^* = N$ and $[(N^* \cdot S)^*]^* = N^* \cdot S$, by 2.24 and 5.25. Hence,

$$(N^* \cdot S) = (N \times S)^*.$$

5.3. ALGORITHMS

Suppose we are given a weakly representative matrix K of a chain-group N on E over R. Then we can describe a number of constructions, based on linear algebra, for investigating the structure of $M(N)$.

We can for example reduce K to diagonal form with respect to a suitable subset S of E. From the rows of K we can then obtain $r(N)$ linearly independent elementary chains of N, and thence $r(N)$ circuits of $M(N)$, by 5.12. Let these circuits be $X_1, X_2, \dots, X_r$, where $r = r(N)$.

Next we can determine the elementary separators of $M(N)$. Starting with X_i, for example, we can take its union with all the other X_j

meeting it, then the union of the resulting set with all the remaining X_k meeting it, and so on, until the process terminates. Let the resulting subset of E be T. Then T is a separator of $M(N)$, by 3.21. It now follows from the construction of T that T is an elementary separator. When we have determined the elementary separators containing the X_i we know that each remaining $a \in E$ determines the elementary separator $\{a\}$. ·

Now let $M(N)$ be binary and let Y be a given circuit of $M(N)$. The chain-group N is not necessarily binary; it may be integral and regular, by 1.26. We construct a weakly representative matrix K of N, diagonal with respect to a subset S of E, such that some row represents a chain f such that $\|f\| = Y$ (see 5.13). Write $S \cap \|f\| = \{a\}$, and $S - \|f\| = S_1$.

Deleting the zero row of $K(E - \|f\|)$ we obtain a weakly representative matrix K_1 of $N \cdot (E - \|f\|)$, diagonal with respect to S_1. The elementary separators of $M[N \cdot (E - \|f\|)] = M(N) \cdot (E - \|f\|)$ can be obtained from K_1. These are the bridges of Y in $M(N)$.

Let B be one of these bridges and let $D(B)$ be the submatrix of K consisting of all the rows having nonzero elements in $K(B)$. We note that if B_1 and B_2 are distinct bridges of Y in $M(N)$, then $D(B_1)$ and $D(B_2)$ are disjoint.

Consider any row of $D(B)$. It corresponds to a circuit X of $M(N)$ and the part of it in K_1 corresponds to a circuit Z of $M(N) \cdot (E - Y)$ on B. Then $Y \cup Z$ is a line of $M(N)$ and the partition of Y determined by Z is $\{X \cap Y, Y - X\}$. Applying this result to all the rows of $D(B)$ we readily deduce that two cells of Y belong to the same member of $\pi[M(N), B, Y]$ if and only if they correspond to equal columns in $D(B)$.

Thus we can determine the bridges of Y in $M(N)$, and the partitions of Y that they determine, from a suitably constructed weakly representative matrix K. Having obtained these partitions we can determine which pairs of bridges overlap and decide whether or not Y is bridge-separable, or totally bridge-separable.

5.4. BRIDGES IN REGULAR MATROIDS

One fundamental result about regular matroids can be expressed as follows.

5.41.   Let P be a plane of a regular matroid M, and let X_1, X_2, and X_3 be points of M on P. Then it is not possible to find seven cells $a_1, a_2, a_3, b_1, b_2, b_3, c$ of P having the following properties:

(i) a_i belongs to X_i but no other member of $\{X_1, X_2, X_3\}$ ($i = 1, 2, 3$).

(ii) b_i belongs to each member of $\{X_1, X_2, X_3\}$ other than X_i, but not to X_i ($i = 1, 2, 3$).

(iii) $c \in X_1 \cap X_2 \cap X_3$.

Proof.   Let us suppose seven such cells exist. Let N be a regular chain-group such that $M = M(N)$. Let f_1, f_2, and f_3 be primitive chains of N with supports X_1, X_2, and X_3, respectively. Multiplying by -1 when necessary we can arrange that $f_i(a_i) = 1$ for each i.

There is a standard representative matrix K of the regular chain-group $N \times P$ whose three rows represent the restrictions to P of f_1, f_2, and f_3. It is associated with the set $S = \{a_1, a_2, a_3\}$. The following diagram shows the columns of K corresponding to the seven cells under consideration:

$$
\begin{array}{c@{\quad}ccccccc}
 & a_1 & a_2 & a_3 & b_1 & b_2 & b_3 & c \\
f_1 & \begin{pmatrix} 1 \\ 0 \\ 0 \end{pmatrix} & 0 & 0 & 0 & x & y & \begin{matrix} z \\ \epsilon z \\ \eta \epsilon z \end{matrix}
\end{array}
$$

$$
\begin{array}{c@{\ }c@{\ }ccccccc}
 & a_1 & a_2 & a_3 & b_1 & b_2 & b_3 & c \\
f_1 & \left(\begin{array}{c} 1 \\ 0 \\ 0 \end{array}\right. & \begin{array}{c} 0 \\ 1 \\ 0 \end{array} & \begin{array}{c} 0 \\ 0 \\ 1 \end{array} & \begin{array}{c} 0 \\ t \\ \eta t \end{array} & \begin{array}{c} x \\ 0 \\ \eta \epsilon x \end{array} & \begin{array}{c} y \\ \epsilon y \\ 0 \end{array} & \left.\begin{array}{c} z \\ \epsilon z \\ \eta \epsilon z \end{array}\right)
\end{array}
$$

The numbers x, y, x, t, η, and ϵ are each equal to $+1$, or -1. We denote the coefficients of b_2, b_3, and c in f_1 by x, y, and z, respectively, and the coefficient of b_1 in f_2 by t. We then write the coefficient of b_3 in f_2 as ϵy. The coefficient λ of c in f_2 is then fixed as ϵz, since

$$
\begin{vmatrix} y & z \\ \epsilon y & \lambda \end{vmatrix} = 1, -1, \text{ or } 0,
$$

by the corollary to 5.21. We next put the coefficient of b_1 in f_3 equal to ηt. Another application of this corollary now shows that the coefficient of c in f_3 is $\eta\epsilon z$, and a third application of it fixes the coefficient of b_2 in f_3 as $\eta\epsilon x$.

But now

$$(0, 1, \text{or} -1) = \begin{vmatrix} 0, & x, & y \\ t, & 0, & \epsilon y \\ \eta t, & \eta\epsilon x, & 0 \end{vmatrix},$$

by the corollary to 5.21,

$$= 2xyt\eta\epsilon,$$

which is absurd. The theorem follows.

5.42.   Let M be a regular matroid on E with a circuit Y. Let Z_1 and Z_2 be distinct points on a connected line L of $M \cdot \bar{Y}$. Then Z_1 and Z_2 are not skew with respect to Y.

Proof. $L \cup Y$ is a plane P of M, by 4.12. There are points X_1 and X_2 of M on it such that $X_1 \cap (P - Y) = Z_1$ and $X_2 \cap (P - Y) = Z_2$. We note that Z_1 determines the partition $\{X_1 \cap Y, Y - X_1\}$ of Y, and analogously for Z_2.

Since L is a connected line we can find $a_1 \in Z_1 - Z_2 = X_1 - (X_2 \cup Y)$, $a_2 \in Z_2 - Z_1 = X_2 - (X_1 \cup Y)$, and $b_3 \in Z_1 \cap Z_2 = (X_1 \cap X_2) - Y$.

If Z_1 and Z_2 are skew with respect to Y we can find also $a_3 \in Y - (X_1 \cup X_2)$, $b_1 \in (X_1 \cap Y) - X_2$, $b_2 \in (X_2 \cap Y) - X_1$, and $c \in X_1 \cap X_2 \cap Y$.

Writing $Y = X_3$ we observe that this result contradicts 5.41. Thus Z_1 and Z_2 are not skew with respect to Y.

5.43.   Let M be a regular matroid on E. Let B be a bridge of a circuit Y of M such that $\pi(M, B, Y)$ has at least two members. Then if $W \in \pi(M, B, Y)$ there exists $Z \in M \cdot \bar{Y}$ determining the partition $\{W, Y - W\}$ of Y.

Proof. Let a be any cell of Y. Let U be a subset of Y such that $a \in U$, some circuit Z of $M \cdot \overline{Y}$ determines the partition $\{U, Y - U\}$ of Y, and $|U|$ has the least value consistent with these conditions.

Suppose there exists $W \in \pi(M, B, Y)$ such that W cuts U. Then there exists a circuit X of $M \times (B \cup Y)$ such that X cuts U. Clearly $X \cup Y$ is a connected flat of $M \times (B \cup Y)$. By 3.42 and 3.43 there is a point X' of $M \times (B \cup Y)$ on $X \cup Y$ that cuts U and is such that $X' \cup Y$ is a connected line of M. Then $X' - Y$ is a point of $M \cdot \overline{Y}$ on B, by 4.12. It determines the partition $\{X' \cap Y, Y - X'\}$ of Y and so severs U.

By 3.43 and 4.31 we can find a point Z' of $M \cdot \overline{Y}$ on B such that Z' severs U and $Z \cup Z'$ is a connected line of $M \cdot \overline{Y}$. By 5.42 Z' determines a partition $\{S, Y - S\}$ of U such that $S \subset U$. By the definition of U we have $a \in U - S$.

Choose $b \in S$. By 4.12 the set $Y \cup Z \cup Z'$ is a plane P of M. The line $\langle P - \{b\}\rangle$ of M is on the points $Z \cup (Y - U)$ and $Z' \cup (Y - S)$ of M. These are distinct and not disjoint. Hence there is a third point T of M on $\langle P - \{b\}\rangle$. By 3.41 this must determine the partition

$$\{U - S, Y - (U - S)\}$$

of Y. But this contradicts the definition of U.

We deduce from this result that U is itself a member of $\pi(M, B, Y)$. Since a may be any cell of Y, the theorem follows.

5.5. TOTALLY BRIDGE-SEPARABLE CIRCUITS

By the pleasing simplicity of graph theory each vertex is normally joined to another by an edge. An analogous statement can be made for regular matroids, but this statement is a rather deep theorem. We take the analogue of a vertex in a regular matroid to be a totally bridge-separable circuit. To justify this we observe that if a nonseparable graph G has no loop then each vertex has a star that is a totally bridge-separable circuit of $B(G)$, by the proof of 4.53.

5.51. Let Y be a circuit of a binary matroid M on E and let a be any cell of Y. Let B be a bridge of Y in M that overlaps another bridge C. Then there exists a circuit Y_1 of M with the following properties:

(i) $a \in Y_1$.

(ii) $Y \cup Y_1$ is a connected line of M.

(iii) Y_1 has a bridge B_1 in M such that B is a proper subset of B_1.

Proof. Either B and C are skew with respect to Y or they are equipartite 3-bridges, by 4.35.

In the first case there exist circuits Z_1 and Z_2 of $M \cdot Y$ on B and C, respectively, that determine partitions $\{S_1, T_1\}$ and $\{S_2, T_2\}$ of Y, respectively, such that $S_1 \cap S_2$, $S_1 \cap T_2$, $T_1 \cap S_2$, and $T_1 \cap T_2$ are all non-null.

We may suppose $a \in S_1 \cap S_2$. Let Y_1 be defined as the circuit $Z_2 \cup S_2$ of M. It satisfies (i) by definition, (ii) by 4.12, and (iii) by 4.61 Corollary, 4.62, and 4.65.

In the second case we write

$$\pi(M, B, Y) = \pi(M, C, Y) = \{S, T, U\},$$

where $a \in S$. By 4.32 there is a circuit Z of $M \cdot \bar{Y}$ on C that determines the partition $\{S, T \cup U\}$ of Y. As before we find that the circuit $S \cup Z$ of M satisfies (i), (ii), and (iii).

5.52. Let Y be a circuit of a binary matroid M on E and let a be any cell of Y. Let B be any bridge of Y in M. Then there exists a totally bridge-separable circuit P of M with the following properties:

(i) $a \in P$.

(ii) There is a bridge C of P in M such that $B \subseteq C$.

Proof. By repeated application of 5.51 there is a circuit Y_k of M with the following properties:

(iii) $a \in Y_k$.

(iv) There is a bridge B_k^1 of Y_k in M such that $B \subseteq B_k^1$, and B_k^1 does not overlap any other bridge of Y_k.

If Y_k is totally bridge-separable the theorem holds. If not, let B_k^2 be one of two overlapping bridges of Y_k in M.

We apply 5.51 with B_k^2 replacing B. We obtain a circuit Y_{k+1} of M with a bridge B_{k+1}^2 containing B_k^2 as a proper subset. Moreover, B_k^1 remains a bridge of Y_{k+1} not overlapping any other, by 4.64 and 4.67.

Continuing in this way we obtain a circuit Y_m of M with the following properties:

(v) $a \in Y_m$.

(vi) $B_k{}^1$ is a bridge of Y_m in M not overlapping any other.

(vii) There is a bridge $B_m{}^2$ of Y_m in M such that $B_k{}^2 \subseteq B_m{}^2$, and $B_m{}^2$ does not overlap any other bridge of Y_m.

If Y_m is not totally bridge-separable we find a bridge $B_m{}^3$ overlapping another and proceed as before. We obtain a circuit Y_n such that $a \in Y_n$. It has bridges $B_k{}^1$, $B_m{}^2$, and $B_n{}^3$ such that no one of these overlaps any bridge of Y_n, other than itself, and such that $B_m{}^3 \subseteq B_n{}^3$. Continuing in this way until the process terminates we obtain the required totally bridge-separable circuit P.

An algorithm for constructing totally bridge-separable circuits is given in "From Matrices to Graphs" (*Can. J. Math.*, 16 (1964), 108–127). It uses the standard representative matrices of the binary chain-group N associated with M. Totally bridge-separable circuits correspond to the "nodal rows" of these matrices.

5.53. Let M be a connected binary matroid. Let Y be a totally bridge-separable circuit of M. Let B be a bridge of Y in M. Let V be a member of $\pi(M, B, Y)$. Then there is a bridge C of Y in M and a $W \in \pi(M, C, Y)$ such that the following conditions hold:

(i) $W \subseteq V$.

(ii) W is cut by no circuit of M.

Proof. We may assume V to be cut by some circuit of M, since otherwise the theorem holds with $C = B$ and $W = V$.

Now V is cut by a circuit X of M such that $X \cup Y$ is a connected line, by 3.42 and 3.43. Then $X - Y$ is a circuit of $M \cdot \overline{Y}$, by 4.12, belonging to a bridge B_1, say. Accordingly, some member W_1 of $\pi(M, B_1, Y)$ cuts V. Since B does not overlap B_1 we may suppose $W_1 \subset V$.

If W_1 is cut by some circuit of M we repeat the argument, with B_1 and W_1 replacing B and V, and so on, until the process terminates.

We now come to the main theorem of this section.

5.54. Let M be a connected regular matroid. Let Y be a totally bridge-separable circuit of M. Let B be a bridge of Y in M. Let V be a member of $\pi(M, B, Y)$. Then there is a second totally bridge-separable circuit Z of M with the following properties:

   (i) $Y \cup Z$ is a connected line of M.

   (ii) $Y \cap Z \subseteq V$.

   (iii) No circuit of M cuts $Y \cap Z$.

Proof. By 5.53 we can find a bridge C of Y in M and a member W of $\pi(M, C, Y)$ such that $W \subseteq V$ and W is cut by no circuit of M.

There is a circuit K of $M \cdot \bar{Y}$ on C determining the partition $\{W, Y - W\}$ of Y, by 5.43. Let X denote the circuit $K \cup W$ of M, which is collinear with Y.

We note that $X \cap Y = W$. Moreover, there is a bridge B_1 of X in M such that $Y - W \subseteq B_1$, by the corollary to 4.61.

Choose $b \in W$. There is a totally bridge-separable circuit Z of M such that $b \in Z$ and such that some bridge B_2 of Z in M contains B_1, by 5.52.

Since $Y - W \subseteq B_1 \subseteq B_2$ we have $Y \cap Z \subseteq W$. Hence $Y \cap Z = W$, since no circuit of M cuts W.

Now $Y - W$ is a flat of $M \cdot B_2$. But if U is any circuit of $M \cdot B_2$ contained in $Y - W$, it contains a circuit of $(M \cdot B_2) \cdot B_1 = M \cdot B_1$. But $Y - W$ is itself a circuit of $[M \cdot (E - X)] \cdot B_1 = M \cdot B_1$, by 4.61. Hence $U = Y - W$, that is, $Y - W$ is a circuit of $M \cdot B_2 = [M \cdot (E - Z)] \cdot B_2$. Accordingly, $Y \cup Z$ is a connected line of M, by 4.12.

The totally bridge-separable circuit Y of M thus fulfills the required conditions.

CHAPTER 6

Duality and Homotopy

6.1. NOTE ON THE PRECEDING CHAPTERS

The foregoing chapters originated in the lecture notes of an intro-
ductory course on the theory of matroids. Naturally some important
parts of the theory of matroids were omitted from this course. The
object of the present supplementary chapter is to give some account
of more advanced work. Proofs of the theorems will be omitted, but
references will be given to the papers in which they are to be found.
The chapter still fails to deal with some important recent develop-
ments such as the theory of transversal matroids.

6.2. DUALITY

The notion of duality for chain-groups, as explained in 1.4, can be
extended to matroids. It was indeed pointed out by Whitney in
[4] that every matroid M has a unique dual matroid M^* on the same
set E, and that $(M*)^* = M$.

Duality for chain-groups is defined in terms of the concept of
orthogonality of chains. Duality for matroids can be defined some-
what analogously in terms of a concept of orthogonality for subsets
of the fundamental set E. Two subsets S and T of E are said to be
orthogonal if the number of their common elements is not one. As a
partial justification of this terminology we observe that two such
subsets can never be the domains of orthogonal chains. Given a
matroid M on E we denote by L the class of all non-null subsets of
E that are orthogonal to all the circuits of M. It is shown in [1], Sec.
2.6, that L satisfies Axiom II. So, by 1.12, the minimal members of
L are the circuits of a matroid on E. This is, by definition, the *dual
matroid* M^* of M.

Duality in matroids relates quite naturally to duality in chain-groups. Thus it is shown in [1] (2.6) that if N and N^* are dual chain-groups, then $M(N)$ and $M(N^*)$ are dual matroids. This leads to the observation that the polygon-matroid and bond-matroid of any graph are dual matroids. This in turn enables us to relate duality in matroids to duality in graphs. Let us describe a matroid as *graphic* if it is the bond-matroid of some graph, and as *cographic* if it is the circuit-matroid of some graph. There is a well-known theorem of Hassler Whitney to the effect that a graph is planar if and only if it has a dual graph. Translated into our terminology this theorem asserts that a graph is planar if and only if its circuit-matroid (as well as its bond-matroid) is graphic. In matroid theory it is convenient to take this as the defining property of a planar graph.

The minors of dual matroids are related by a very simple rule. The minors of M^* are the duals of the minors of M.

6.3. HOMOTOPY

Let us consider the problem of characterizing a regular matroid in terms of its minors. We suppose given a binary matroid M. If any minor of M is not regular, then M itself is not regular, by 2.36. It seems desirable therefore to determine all binary matroids M with the following property: M is not regular, but every other minor of M is regular. Let us refer to such matroids as "minimal irregular binary matroids".

The solution to this problem of classification is known. There are just two essentially distinct minimal irregular binary matroids. They can be specified by the following standard representative matrices over $GF(2)$.

$$(1) \quad \begin{pmatrix} 1 & 0 & 0 & 1 & 0 & 1 & 1 \\ 0 & 1 & 0 & 1 & 1 & 0 & 1 \\ 0 & 0 & 1 & 0 & 1 & 1 & 1 \end{pmatrix}$$

$$(2) \quad \begin{pmatrix} 1 & 1 & 0 & 1 & 0 & 0 & 0 \\ 0 & 1 & 1 & 0 & 1 & 0 & 0 \\ 1 & 0 & 1 & 0 & 0 & 1 & 0 \\ 1 & 1 & 1 & 0 & 0 & 0 & 1 \end{pmatrix}$$

These matrices represent dual binary chain-groups and may therefore be said to represent the corresponding dual binary matroids. A matroid corresponding to the first matrix is said to be of Type *BI*, and one corresponding to the second is said to be of Type *BII*.

Our theorem 5.41 can be regarded as a proof that a matroid of Type *BI* is not regular. When this fact is established it is easy to show that the matroid is a minimal irregular binary matroid. The result can then be extended to the matroid of Type *BII* by duality.

The great difficulty in this theory is that of proving that the matroids of Types *BI* and *BII* are the only minimal irregular binary matroids. In what follows we try to explain how this difficulty can be overcome without actually going into the details of the proof. Detailed proofs can be found in References [1] and [2]. We note that the result to be proved is equivalent to the following characterization theorem.

6.31. A binary matroid is regular if and only if it has no minor of Type *BI* or *BII*.

In attempting to prove this result we naturally suppose given a minimal irregular binary matroid M, not of Type *BI* or *BII*, and investigate its properties. We readily show that M must be connected, for otherwise the components would be regular and therefore M would be regular. It is also a simple matter to establish that M must have at least 7 cells and that its rank must be at least 3.

Now M is the matroid $M(N)$ of a chain-group N over $GF(2)$. To prove it regular (and thereby obtain a contradiction) we must show that also $M = M(N_r)$, where N_r is a regular chain-group. We recall that the elementary chains of a regular chain-group are the non-zero integral multiples of the primitive chains, and the coefficients in the primitive chains are restricted to the values 1, -1 and 0. From a

primitive chain of N_r we obtain an elementary chain of N by replacing each coefficient by its residue class mod 2. Conversely from
any elementary chain of N we may expect to obtain a primitive
chain of N_r by "signing" it, that is by replacing each non-zero coefficient by one of the integers 1 and –1. The signing has to be carried
out in such a manner as to preserve linear dependence of elementary
chains. It is fairly easy to show that M is regular if N can be signed in
this way.

We therefore attempt to sign N. We select a cell a and write
$M' = M \cdot (E - \{a\})$. Then M' is binary, being the matroid of the chain-
group $N' = N \cdot (E - \{a\})$. By the minimality of M we can assume that
M' is regular. We can therefore sign the elementary chains of N' so
that linear dependence is preserved.

The elementary chains of N' can be classified in two families A
and B. The members of A are restrictions of chains of N in which a
has a zero coefficient, and the members of B are restrictions of chains
of N in which a has a non-zero coefficient. It is easy to see that the
members of B must be restrictions of elementary chains of N. We try
to sign the extensions of the members of B in N in the following way.
If K is a member of B we first sign K as an elementary chain of N'
Then we choose an integer $s(K, a)$, equal to 1 or –1, to be the corresponding coefficient of a.

The choices of the integers $s(K, a)$ are not all arbitrary. Suppose
for example we have two distinct elementary chains K and L of N'
and that their domains are two points of a connected line of M'.
Denote this line by U. Let b be a cell common to the domains of K
and L. Then if K and L are in B it is necessary to choose $s(K, a)$ and
$s(L, a)$ so that

$$s(K, a) \, s(L, a) \;=\; \overline{K}(b) \, \overline{L}(b) \,,$$

where $\overline{K}(b)$ for example is the integer 1 or –1 which replaces $K(b)$ in
the signing of K. For unless this equation is satisfied the signing
converts $N \times (U \cup \{a\})$ possibly into an integral chain-group but cer-
tainly not into a regular one, by 5.21.

Let us now consider the structure of the matroid M'. The circuits
that are the domains of the members of A constitute a linear subclass

C of M'. A *path* in M' is a finite sequence of circuits of M', not necessarily all distinct, such that any two consecutive terms of the sequence are distinct circuits having as their union a connected line of M'. The path is said to be "off C" if it has no term that is a member of C.

Consider a path $(X_1, X_2, X_3, \ldots, X_n)$ of M', off C. Suppose we have chosen a number $s(K, a)$ for a chain K with domain X_1. Then the corresponding number for X_2 is fixed, by the preceding argument, and so is the number for X_3, and so on. Proceeding all along the path we eventually fix the number for X_n. A case of great interest arises when the path is re-entrant, that is when $X_1 = X_n$. The final number for X_n must agree with the initial choice of $s(K, a)$; otherwise our attempt to derive a signing of N from the given signing of N' will have failed. We therefore require the re-entrant paths of M', off C, to satisfy certain consistency conditions, expressible in terms of the signing of N'.

Let us now state that these consistency conditions are in fact satisfied, and that because of this it is possible to sign N and prove that M is regular, so obtaining the desired contradiction. In this way Theorem 6.31 can be proved. But the proof of the consistency conditions is difficult, and it seems to require the development of a new branch of the theory of matroids, a theory of "homotopy". There is a "Homotopy Theorem" implying that the satisfaction of the consistency conditions for re-entrant paths off C of a few simple kinds is sufficient; if the conditions hold for these simple paths they hold for all re-entrant paths off C. It is in fact easy to show that the "simple" paths occurring in M' do indeed satisfy the consistency conditions.

It remains to explain the meaning of the Homotopy Theorem, and to give some account of its proof. In Reference 2, Paper I is concerned with the proof of the Homotopy Theorem, and Paper II with its application to the characterization of regular matroids.

6.4. THE HOMOTOPY THEOREM

We suppose given a matroid M, not necessarily binary, and a linear subclass C of M. We study re-entrant paths off C. Suppose we have two such paths, P and Q, of the following forms.

$$P = (X_1, X_2, X_3, \ldots, X_n, X_1),$$

$$Q = (X_i, Y_1, Y_2, \ldots, Y_k, X_i),$$

where $1 \leq i \leq n$. Then another re-entrant path off C is

$$R = (X_1, X_2, \ldots, X_{i-1}, X_i, Y_1, Y_2, \ldots, Y_k, X_i, X_{i+1}, \ldots, X_n, X_1).$$

We say that P is *deformed* into R by the adjunction of Q, or that R is *deformed* into P by the deletion of Q. We may also say that R is *decomposable* into P and Q. In homotopy theory we specify a class U of re-entrant paths off C called "elementary". Two paths P and R are then said to be homotopic if P can be transformed into R by a finite sequence of operations, each of which adjoins or deletes an elementary path. We recognize a *null* path, which is of course homotopic to each elementary path. The relation of being homotopic is clearly an equivalence relation. If P and R are homotopic we write $P \sim R$. A path homotopic to the null path is said to be null-homotopic. If P is such a path we write $P \sim 0$.

The problem of homotopy first arose in the following form: can we choose U in some simple way so as to make all re-entrant paths off C null-homotopic? If so the application to the matroid M' of Sec. 6.3 is clear: we must show first that the members of U satisfy the consistency conditions, and secondly that deformation by the adjunction or deletion of a member of U converts a consistent path into a consistent path.

In choosing U it seemed natural to include all paths off C of the following forms: (X, Y, X) on a line and (X, Y, Z, X) on a plane. These are the *elementary paths* of the first and second kinds respectively. Attention was then drawn to paths, off C, on a plane P of the form (X, Y, Z, T, X), where X, Y, Z and T are distinct, there are two distinct points A and B on P such that each connected line on P is on either A or B, $X \cup Y$ and $Z \cup T$ are lines on A, and $Y \cup Z$ and $T \cup X$ are lines on B. It was found to be impossible to transform such a path into the null path by adjoining and deleting elementary paths of the first and second kinds. Such paths were therefore included in U as the *elementary paths of the third kind.*

An attempt was next made to show that a re-entrant path off C, confided to a flat of M of dimension d, (that is rank $d + 1$), could always be deformed into a path in a flat of lower dimension by adjoining and deleting elementary paths of the first, second and third kinds. This attempt succeeded only partially. It was found that the operation is possible for all $d \geq 0$ if it is possible for the case $d = 3$. But a close investigation of the three-dimensional case disclosed a class of paths not deformable into the null path by adjoining and deleting elementary paths already recognized. These paths were therefore included in U as elementary paths of the fourth kind. It could then be shown that, with respect to U, all re-entrant paths off C were null-homotopic. This is the result that we have referred to as the Homotopy Theorem.

6.5. GRAPHIC MATROIDS

In Reference [3] the main problem is that of finding a characterization of graphic matroids analogous to the characterization of regular matroids already discussed. We know of course that every graphic matroid is regular (1.34) and that every minor of a graphic matroid is graphic (2.321 and 2.322). Our problem thus reduces to that of classifying the matroids M with the following property: M is regular and non-graphic, but every other minor of M is graphic. Such matroids we call the minimal non-graphic regular matroids. Evidently a regular matroid is graphic if and only if it has no minor that is a minimal non-graphic regular matroid.

It is known that there are essentially only two such minimal matroids. They are the polygon-matroids of the Kuratowski graphs. The method of proof of this statement can be described very briefly as follows. We take a proof of Kuratowski's Theorem and generalize it from graphs to regular matroids. In the appropriate generalization planar graphs are replaced by graphic matroids. Use is made of the theory of regular matroids presented in Chapter 5.

6.6. THREE-CONNECTED MATROIDS

There are several slightly different definitions of connectivity for graphs. The one relevant here runs as follows.

If S and T are complementary subsets of $E(G)$ we write $\eta(G; S, T)$ for the number of common vertices of $G \cdot S$ and $G \cdot T$. We say G is *k-separated*, where k is a positive integer, if G is connected and there are complementary subsets S and T of $E(G)$ such that

$$\eta(G; S, T) = k,$$

$$\text{Min}(|S|, |T|) \geq k.$$

We say G is *0-separated* if and only if it is not connected. If there is a least non-negative integer k such that G is k-separated we call it the *connectivity* of G and denote it by $\lambda(G)$. If there is no such integer we write $\lambda(G) = \infty$.

An attempt to extend the notion of connectivity to matroids led to the following definition.

A matroid M is said to be *k-separated*, where k is a positive integer, if there are complementary non-null subsets S and T of E such that

$$r(M) - r(M \times S) - r(M \times T) + 1 = k,$$

$$\text{Min}(|S|, |T|) \geq k.$$

If there is a least positive integer k such that M is k-separated, we call it the *connectivity* of M and denote it by $\lambda(M)$. If there is no such integer we write $\lambda(M) = \infty$.

It can be shown that the connectivity of a connected graph is equal to the connectivity of its polygon-matroid. The proof is surprisingly difficult. It can be shown also that the connectivity of a matroid is equal to the connectivity of its dual matroid. For the full theory reference may be made to the author's paper "Connectivity in Matroids" (*Can. J. Math.*, 18 (1966), 1301–1324).

It is well-known that if $\lambda(G) \geq 3$, that is G is "three-connected", then either G is a wheel or it can be transformed into a smaller three-connected graph by deleting or contracting a single edge. In the paper just mentioned an analogous theorem is proved for matroids. It states that if M is a non-null matroid such that $\lambda(M) \geq 3$, then either M is a "wheel" or a "whirl" or there is a cell a such that either

$M \cdot (E - \{a\})$ or $M \times (E - \{a\})$ has connectivity at least 3. Here a "wheel" is the polygon-matroid of a wheel in the graphic sense. A "whirl" is derived from such a wheel by deleting the rim from the list of circuits but recognizing any set formed by adjoining a single spoke to the rim as a new circuit.

REFERENCES

1. Tutte, W. T., "Lectures on Matroids," *J. Res. Nat. Bur. Std., B,* 69 (1965), 1–47.
2. Tutte, W. T., "A Homotopy Theorem for Matroids," I, II, *Trans. Amer. Math. Soc.,* 88 (1958), 144–174.
3. Tutte, W. T., "Matroids and Graphs," *Trans. Amer. Math. Soc.,* 90 (1959), 527–552.
4. Whitney, H., "On the Abstract Properties of Linear Dependence," *Amer. J. Math.,* 57 (1935), 507–533.

Index